VIDEOPHILOSOPHY

COLUMBIA THEMES IN PHILOSOPHY,
SOCIAL CRITICISM, AND THE ARTS

COLUMBIA THEMES IN PHILOSOPHY, SOCIAL CRITICISM, AND THE ARTS

LYDIA GOEHR AND GREGG M. HOROWITZ, EDITORS

Advisory Board

CAROLYN ABBATE	ARTHUR C. DANTO
J. M. BERNSTEIN	JOHN HYMAN
EVE BLAU	MICHAEL KELLY
T. J. CLARK	PAUL KOTTMAN

Columbia Themes in Philosophy, Social Criticism, and the Arts presents monographs, essay collections, and short books on philosophy and aesthetic theory. It aims to publish books that show the ability of the arts to stimulate critical reflection on modern and contemporary social, political, and cultural life. Art is not now, if it ever was, a realm of human activity independent of the complex realities of social organization and change, political authority and antagonism, cultural domination and resistance. The possibilities of critical thought embedded in the arts are most fruitfully expressed when addressed to readers across the various fields of social and humanistic inquiry. The idea of philosophy in the series title ought to be understood, therefore, to embrace forms of discussion that begin where mere academic expertise exhausts itself; where the rules of social, political, and cultural practice are both affirmed and challenged; and where new thinking takes place. The series does not privilege any particular art, nor does it ask for the arts to be mutually isolated. The series encourages writing from the many fields of thoughtful and critical inquiry.

Lydia Goehr and Daniel Herwitz, eds., *The Don Giovanni Moment: Essays on the Legacy of an Opera*

Robert Hullot-Kentor, *Things Beyond Resemblance: Collected Essays on Theodor W. Adorno*

Gianni Vattimo, *Art's Claim to Truth*, edited by Santiago Zabala, translated by Luca D'Isanto

For a complete list of titles see pages 277–278

VIDEOPHILOSOPHY

THE PERCEPTION OF TIME IN POST-FORDISM

MAURIZIO LAZZARATO

EDITED AND TRANSLATED BY
JAY HETRICK

Columbia University Press   *New York*

Columbia University Press
Publishers Since 1893
New York   Chichester, West Sussex
cup.columbia.edu

Copyright © 2019 Columbia University Press

All rights reserved

Library of Congress Cataloging-in-Publication Data
Names: Lazzarato, M. (Maurizio), author.
Title: Videophilosophy : the perception of time in
post-Fordism / Maurizio Lazzarato ; edited and translated
by Jay Hetrick.
Other titles: Videofilosofia. English
Description: New York : Columbia University Press, 2019. | Series: Columbia themes in philosophy, social criticism, and the arts | Includes bibliographical references and index.
Identifiers: LCCN 2018027649 | ISBN 9780231175388 (cloth) |
ISBN 9780231175395 (pbk.) | ISBN 9780231540162 (e-book) Subjects: LCSH:
Aesthetics—Political aspects. | Bergson, Henri, 1859–1941. | Time perception.
Classification: LCC BH301.P64 L3913 2019 | DDC 111/.85—dc23
LCrecordavailableathttps: //lccn.loc.gov/2018027649

Cover design: Lisa Hamm

Cover image: Video still from Angela Melitopoulos and Maurizio Lazzarato,
The Life of Particles, 2013. Used with permission of the artists.

To Angela and the crystals of time of her videos

CONTENTS

Lazzarato's Political Onto-aesthetics ix
JAY HETRICK

Introduction 1

1 The War Machine of the Kino-Eye and the Kinoki Against the Spectacle 19

2 Bergson and Machines That Crystallize Time 37

3 Video, Flows, and Real Time 81

4 Bergson and Synthetic Images 109

5 Nietzsche and Technologies of Simulation 139

6 The Economy of Affective Forces 169

7 The Concept of Collective Perception 199

Afterword: Videophilosophy Now—an Interview with Maurizio Lazzarato 227

Notes 239
Index 263

LAZZARATO'S POLITICAL ONTO-AESTHETICS

JAY HETRICK

Maurizio Lazzarato originally wrote *Videophilosophy* in 1996 as his doctoral thesis under the supervision of the Marxist philosopher, and director of the Department of Political Science at the University of Paris 8, Jean-Marie Vincent. The original title was "Machines that Crystallize Time: Perception and Labor in Post-Fordism." The manuscript was then rewritten and published in Italian as *Videophilosophy: The Perception of Time in Post-Fordism* in 1997, the same year as his *Immaterial Labor: Forms of Life and the Production of Subjectivity*.[1] Of course, Lazzarato is best known internationally for his concept of immaterial labor. However, while this latter book is more a treatise on contemporary political economy, *Videophilosophy* is a book of speculative philosophy that can be understood as providing the conceptual ground for Lazzarato's work as a whole and, especially, for his recent forays into the politics of art. As such, *Videophilosophy* lies at the heart of Lazzarato's conceptual apparatus and is therefore the book that most clearly distinguishes him philosophically from his peers. His singular intervention into the recent conversations concerning post-Marxist political ontology as well as the politics of aesthetics is ultimately connected to his reliance on Henri Bergson and

Friedrich Nietzsche—but also Walter Benjamin, Gilles Deleuze, Michel Foucault, Félix Guattari, and Gabriel Tarde—in order to construct an entirely different non-Hegelian and post-Marxist metaphysics of contemporary capitalism. Lazzarato doesn't mince his words on this point: "While Marx indicated the methodology with which to discover 'living labor' beyond work, he is of no help in analyzing . . . the conditions of contemporary capitalism." Instead, and quite remarkably, it is "Bergson and Nietzsche [who] should be understood as the conceptual personae who have constructed an ontology for contemporary capitalism."[2] *Videophilosophy* will appeal to film and media theorists, since it builds upon both Deleuze's *Cinema* books and Benjamin's essay "The Work of Art in the Age of Its Technological Reproducibility" in interesting and novel ways. But *Videophilosophy* is also, and perhaps more profoundly, an original and compelling work of political ontology, as long as we understand the political primarily on the level of micropolitics or the microphysics of power. Furthermore, Lazzarato follows Deleuze and Guattari—especially their conceptions of signs and asignifying semiotics, respectively—in intertwining the ontological and aesthetic registers. That is, *Videophilosophy* does not simply offer a philosophy of video art—an idea consistent with the book's original title—but is also a work of political onto-aesthetics.

Lazzarato's analyses of the nature of post-Fordist capitalism—first using the concept of immaterial labor and more recently in his books on debt—hinge on the fact that the field of political economy has radically shifted since the early 1970s. Despite the more problematic aspects of Lazzarato's theory of labor—which he has largely moved away from since commencing his work on Tarde in the early 2000s—his overall point of view with respect to contemporary capitalism has remained the same.[3] For him, capitalism is no longer simply about modes of production and consumption but has primarily become an apparatus for "machinic enslavement" that operates by capturing and controlling the precognitive and even preindividual elements of subjectivity: constantly shifting assemblages of unindividuated affects, percepts, and, more generally, what Guattari calls asignifying signs. For Lazzarato, money in the time of neoliberalism also functions on this asignifying plane, which is why it can now be understood as "a political apparatus."[4] Furthermore, capitalism's increasing reliance on the information, service, and attention economies goes hand in hand with its exploitation of the elements of this plane, aided by the new technologies associated with post-Fordism. This plane is populated by what Lazzarato calls "the genetic, creative, differential element that Marx defines as living labor," which lies beneath traditional categories of political economy such as factory discipline and wage relations.[5] He boldly argues that in order to analyze the genetic elements of this plane properly, we must look at the process of the production of subjectivity, since it is more relevant in this regard than the concepts typically used in the critique of political economy.

Therefore, as a part of Lazzarato's larger critique of strategies of resistance, the old idea of constructing a political program must necessarily be preceded by ethical and microsocial endeavors concerned with the continuous experimental renegotiation of both subjectivity and the relationship between the individual and the collective. With this crucial point in mind, the uncritical accusations of vitalism against Lazzarato and his philosophical lineage appear entirely misguided, since his use of Bergson, Nietzsche, and Tarde is aimed precisely at this micropolitical level. "Micropolitics is far from being a call to spontaneity, a simple call to movement, a simple affirmation of forms of life (a vitalism as Jacques Rancière or Alain Badiou would say with disdain). Micropolitics requires a very high level of organization, a precise differentiation of the actions and the functions of the political, a multiplicity of initiatives, an intellectual and organizational discipline."[6] Charges of "Bergsonist!"[7] against Lazzarato cannot mean the same thing as they have when applied to thinkers like Antonio Gramsci or Georges Sorel, since Lazzarato is not primarily attempting to construct a political program, "spontaneous" or otherwise.[8] In fact, Bergson's appropriation and simultaneous vilification by critical theorists is an area of research that is largely unwritten even though it has been entirely overcoded by György Lukács's disingenuous attack on vitalism in his 1952 book *The Destruction of Reason*, where he dismisses vitalism as a kind of proto-Fascist thinking. The historical truth, as pointed out by both François Azouvi and Carl Schmitt, is that Bergson has been deployed by both the nationalist right and the anarchist left.[9] Furthermore, the concept of vitalism still needs to be fully and honestly analyzed[10]—rather than thrown around as a philosophical straw man—and the subterranean use of Bergson by critical theorists including Theodor Adorno, Benjamin, Ernst Bloch, Max Horkheimer, and Lukács himself needs to be fully assessed. Finally, one of Lazzarato's unspoken premises seems to be that a dialectical logic based on Deleuzo-Guattarian readings of Bergson, Nietzsche, and Tarde can allow concepts like conflict and antagonism to subsist in an environment stripped of negative dialectics.

Besides the concept of post-Fordism, all the key words from the original title of *Videophilosophy*—machines, crystal, time, perception, and even labor—refer directly to Lazzarato's reading of Bergson, which itself is mediated by Deleuze's two books on cinema. In these books, Deleuze famously claims that the universe is a "metacinema." He makes this claim with a Bergsonism that has been operated upon by Nietzsche and Gottfried Leibniz in order to (1) disallow a conception of "pure perception" that might lay claim to any concept of truth and (2) to ensure that this universe—sliced up by framing, shooting, and montage—is understood as radically acentered and discontinuous.[11] This is, as Anne Sauvagnargues notes, a "cinemachinic universe."[12] I will come back to the important concept of the machine later but will simply note now that, along with the "inorganic" concept of the crystal, it is part of Deleuze's critique of phenomenological perception. It also prevents a collapse into some kind of mystical vitalism and, as a concept that supersedes techne, blurs the boundary between nature and artifice. While the temporal metaphysics Deleuze constructs in his *Cinema* books is grounded in the philosophies of Bergson, Leibniz, and Nietzsche, Lazzarato's metaphysics of contemporary capitalism is similarly based on the work of Bergson, Nietzsche, and the Leibnizian sociologist Tarde. Therefore, Lazzarato also places the problem of time at the heart of his philosophy. In this, albeit in an entirely different way, he follows his comrades Antonio Negri and Éric Alliez.[13] In any case, if we understand Deleuze's *Cinema 1* and *Cinema 2*, reductively, as presenting a historical progression from pre- to postwar cinema and philosophy, Lazzarato's *Videophilosophy* could be seen—in an equally reductive way—as something like *Cinema 3*, since from the same metaphysical ground, it announces a shift from cinema to video art in the 1970s. Furthermore, just as World War II was seen by Deleuze and others as an event that completely altered the landscape of possibilities and impossibilities for art and thought, the Nixon shock in the early 1970s was for Lazzarato an event that has completely transformed our political, social, and ethico-aesthetic realities.[14]

More important than the shift from cinema to video and electronic media—whose real differences for Lazzarato lie not in the concept of indexicality but rather in each medium's ability to crystallize time and ultimately express machinic forces—is the fact that, with neoliberalism, we have entered a new era socially, politically, and economically. Therefore, Lazzarato takes care to historicize and politicize his conceptual personae. Leibniz, as I have already mentioned, is ultimately replaced by Tarde, while Nietzsche becomes the grand theorist of debt in Lazzarato's most recent work. But even in *Videophilosophy*, Lazzarato pushes Bergson as far into the political domain as possible. In order to remove any inkling of ambiguity, any possibility of understanding these ideas as "out of this world," he goes beyond Deleuze by clearly stating that "I cannot follow the path that led Bergson toward a new spirituality."[15] Lazzarato uses Bergson primarily to construct an "ontology of the new economy" and an "ontology for a definition of the society of the image," which for him amount to the same thing.[16] For example, he alludes to a consonance—while being careful not to conflate them entirely—between his own concept of intellectual labor and the type of intellectual effort involved in the Bergsonian theory of perception and production of images, which Lazzarato will refer to as the crystallization of time: "The theory of the production of images in Bergson is not an optical, but rather a temporal, theory that can be explained by the different forms of contraction-relaxation of time; namely, the different syntheses of intellectual labor."[17] Lazzarato also finds Bergson's call for a second industrial revolution in *Two Sources of Morality and Religion* useful and "quite original." Interestingly, in 1935, a year after Max Horkheimer's critique of Bergson, Bloch found these passages promising as well: "There is no longer the slightest *anti-intellectual* romanticism or irrationality of life per se, as in the former 'cosmic' Bergson. . . . The creator of the philosophy of life is no stranger to the courage of the most advanced technology, indeed he even aims . . . at an equally anti-individual and anti-national, planned economy."[18] But beyond these few small gestures, it is really with the aid of Benjamin and Dziga Vertov that Bergsonism is made historical and political.

Lazzarato turns to Benjamin's concept of collective perception specifically to read Bergson on a more political ground and not to "move beyond" him, as Alberto Toscano has suggested.[19] In fact, Lazzarato explicitly says it is Bergson who "allows us to move beyond" Benjamin.[20] It is Bergson who gives us a more convincing account of the processes of image production in the time of contemporary capitalism. Furthermore, in a Bergsonian light Benjamin's concepts of technological reproduction and *Jetztzeit* become problematic and in need of being reconceived. The former is replaced by the concept of machinic apparatuses, which I will discuss later, and the latter is supplemented by Bergson's concept of virtual memory, from his book *Matter and Memory*. Perhaps Lazzarato mentions Benjamin's analysis of that book in "On Some Motifs in Baudelaire" in order to hint that Benjamin is closer to his own lineage, and therefore further away from the negative dialectics of Adorno and Horkheimer, than is usually acknowledged. This has been corroborated by Axel Honneth, the current director of the Institute for Social Research, who claims that "his concern with Bergson's writings on the philosophy of life . . . enabled Benjamin to give his ideas about non-mechanical, richly meaningful experience clear contours."[21] In any case, Lazzarato turns to Benjamin because he connects the mechanization and collectivization of labor to the mechanization and collectivization of perception. He also associates modern forms of socialization with the birth of cinema, as evidenced by the fact that there emerges simultaneously a "shock produced by the assembly line and by montaged images."[22] Finally and crucially, Benjamin notices that with the advent of cinema, we see a "flattening" of the separation between intellectual and manual labor as well as a "reversibility" between producer and consumer, both of which anticipate the transformations of subjectivity specific to post-Fordist capitalism.

Lazzarato's machinic, or crystallized, articulation of Bergsonian intuition is made revolutionary by his conflating it with the flash of recognition involved in Benjamin's theory of Jetztzeit. Interestingly, Benjamin uses similar, albeit politicized, language to describe this experience: "Where thinking suddenly comes to a stop in a constellation saturated with tensions, it gives that constellation a shock, by which it is crystallized... and can be seized only as an image that flashes up at the moment of its recognizability... the sign of a messianic arrest of happening, or (to put it differently) a revolutionary chance in the fight for the oppressed past."[23] For Lazzarato, this flash is understood as a moment in which the sensory-motor habits of capitalism are ruptured and the empty and homogeneous continuity of "value-time" is replaced by a more primary nonchronological "power-time." This is the time of invention, not simply of commodities, but of new worlds; that is, new percepts, affects, and beliefs. Lazzarato argues that the ambiguity of the concept of Jetztzeit is due to the fact Benjamin failed to fully articulate historical time—which ultimately determines the conditions for a political awakening of the image—in its ontological, or messianic, form. Lazzarato reads the messianic present—which Benjamin posed as an alternative to both the measured time of capital and the impossible return to the time of tradition—through a decidedly Bergsonian lens as the time that contains all times, or the virtual past. This moment is conceived as potentially revolutionary, since an expanded collective perception, understood as the machinic perception of power-time on a mass scale, could ultimately lead to an innervated form of collective action.

Lazzarato inherits this idea of short-circuiting the habitual patterns of the sensory-motor system from Deleuze's *Cinema* books, which are also the key to understanding his concept of "machines that crystallize time." For Deleuze, the moment of short-circuiting—induced by the pure sonic and optical percepts expressed in particular films—causes psychic energy to collect within the brain, making it difficult to react immediately to a given situation. This type of picnoleptic lapse in consciousness should not be interpreted as necessarily resulting in pacifism or quietism—as Peter Hallward's critique of Deleuze suggests[24]—but as a contemporary phenomenon that forces us to rethink the very categories of labor and action, which Lazzarato does with the help of not only Bergson but also Marcel Duchamp.[25] In any case, while Deleuze understands the work of Dziga Vertov as fulfilling the metaphysics of Bergson's *Matter and Memory* cinematically, he fails to emphasize Vertov's political commitment. For him, Vertov's cinematic vision realizes the genetic element of the perception-image—referred to as "gaseous perception" by Deleuze—which defines the very nature of the cinemachinic universe. Here, perception is extended so far beyond the human that it potentially reaches "the *clinamen* of Epicurean materialism."[26] This gaseous perception is the *kino-eye*, the Vertovian theory of an eye in matter that is able to connect any point in the universe to any other and in any temporal order. Lazzarato appropriates this onto-aesthetic understanding of the kino-eye.[27] Beyond Deleuze, he does so in order to construct a political ontology, such that the moment of short-circuiting induced by cinematic machines becomes the time of the event—in the sense that a situation's cohesion becomes temporarily disrupted, forcing us to invent new modes of being in the world. This is why, for Lazzarato, cinema and video are understood as "war machines" that can be directed against our contemporary society of the spectacle. He follows Vertov in conceiving cinesensibility as "a major political issue," since it is a "powerful tool for the repudiation of the imperialism of signifying semiotics."[28] "The kino-eye is a machine for the contraction-relaxation; that is, the crystallization of time.... Cinema is the first of a new type of machine that fixes and reproduces the time of perception, sensibility, and thought [beyond the human coordinates] ... rendering sensible new matter, new affects, and new forces."[29] The film theory and practice of Vertov therefore become a kind of methodological framework for *Videophilosophy*, offering a concrete example of how film, and video after it, is fundamentally intertwined with an ontological, aesthetic, and most importantly, political vision. With this politicized Bergsonism, we can now turn to Lazzarato's central concept of machines that crystallize time. But I should first note that while the flash moment that wrenches our habitual sensibilities beyond recognition is understood as the time of the event, Lazzarato also acknowledges a more primary form of the event for Deleuze—*eventum tantum*—that is inherent to the discontinuous flow of matter itself as a bricolage of aberrant intervals (jumps, cuts, rhythms) that form the "the nonimaged ground, the deterritorialized flux from which images emerge."[30] These onto-aesthetic events are, in Deleuze's reading of Leibniz, both everywhere and remarkable, since they form the ruptured fabric of time-matter itself. For Lazzarato, they are the ontological basis for remarkable ruptures in the ethical, social, and political domains.

In Lazzarato's reading, the ultimate purpose of the kino-eye is to make us see, because, to riff on Martin Heidegger, we are not yet seeing. Film and video technologies give us access to the cinemachinic universe of images—or, more precisely, the temporal universe of intervals and signs—and therefore facilitate the renegotiation of subjectivity. But in the time of contemporary machines, the cinematic "seer" is pushed beyond the idea of a mystic visionary and even beyond phenomenological perception itself. Here, seeing becomes the crystallization of time. To stress this nonphenomenological approach, Lazzarato deploys the word *crystallization* to describe the Bergsonian contraction and relaxation of time-matter that produce images. The cinemachinic universe is nothing but a vortex of images that encounter and collide with one another. All images, including human subjects, are assembled contractions and relaxations of time-matter. Like the human body, technologies of photography, cinema, video, and electronic media function as special types of images that are increasingly capable of accumulating and expressing the power of time, which in turn allows for ever greater possibilities of action, invention, and the construction of subjectivity more generally. As Lazzarato notes, this point of view is diametrically opposed to that of Paul Virilio, for whom new technologies inhibit our access to real time and, therefore, diminish our capacity to act. In the time of immaterial labor, the old forms of exploitation and disciplinary subjection have been displaced by what Lazzarato calls processes of machinic enslavement, which capitalism uses to manipulate subjectivity on a subrepresentational level. The mass media—and particularly television, understood primarily as an apparatus of power even though it has access to real time—has appropriated the ability of new technologies to capture and overcode not only the plane of what Guattari calls signifying signs (the recognizable world of images and words) but also the plane of asignifying signs, which includes "mathematics, stock quotes, money, business and national accounting, computer languages, the functions and equations of science; but also the semiotics of music and art."[31] A crucial political question in this control society therefore becomes, how can we develop practices that might intervene and *détourner* this situation by reclaiming the power of machines that crystallize time?

Lazzarato roughly follows Deleuze's Bergsonian film philosophy by arguing that cinema reveals the world as a flow of images.[32] However, for Lazzarato the genetic element of cinema is still the photograph. And while montage adds a temporal element, "it does not yet extract from the infinite variation of asignifying figures, and it does not yet plunge into image-matter."[33] Although film does not express the variety of asignifying signs associated with the electronic deconstruction of the image—into what Bergson calls "visual dust"[34]—it is still a complex assemblage that offers the possibility of plugging into multiple semiotic registers simultaneously: "images, sounds, words spoken and written (subtitling), movements, postures, colors, rhythms."[35] While the film industry has of course learned how to manipulate and capitalize on this motley assemblage of different signs, Lazzarato, following Guattari, believes that ultimately these cinematic signs cannot be completely policed and overcoded. Some nonrecuperable excess remains, which can help "produce desubjectification and disindividuation effects . . . and it can strip the subject of his identity and social functions."[36] Beyond this, Lazzarato then presents an entire taxonomy of signs that we encounter in video art, which should be understood as adding to the intrinsic qualities of cinema. Video technology enables a further deterritorialization of these flows by expressing not only images in movement but also the very conditions of the image, the time-matter of electromagnetic waves that lies at the heart of both the video image and the physical world itself: "Video technology is a *machinic assemblage that establishes a relationship between asignifying flows (waves) and signifying flows (images)*. It is the first technical means of image production that corresponds to the generalized decoding of flows."[37] Instead of simply utilizing words, symbols, or images, video acts as a kind of electronic paintbrush to construct and express, beyond signification, point-signs, which are themselves the genetic conditions of words and images. That is, rather than capture images, the video camera captures the waves that constitute images, composing and decomposing them by means of crystallization or, to speak with Gilbert Simondon, modulation. By drawing on the writings of several video artists, including Nam June Paik, Bill Viola, and Angela Melitopoulos, Lazzarato argues that video is a machine that is able to crystallize and express the multitudes of intensities and rhythms that are ontologically prior to all signification and representation. Therein lies its power to create the new, to encourage unheard-of affects and bodily pulsations, and to proliferate beyond the controlling power of television, which though also able to access the asignifying world, does so only to capture and mold these asignifying elements into given forms of subjectivity.

Remarkably, therefore, the Bergsonian image is realized through technology. But it is important to note that technological machines—"the mechanist vision of the machine"—are simply one type of machine, which should be understood as a much broader category.[38] Here, Lazzarato appropriates Guattari's concept of the machine, which includes social, economic, aesthetic, linguistic, biological, cosmic, and ecological machines, as well as the type of abstract machine Guattari conceptualized together with Gilles Deleuze. Guattari's position is that in the face of new ecological challenges brought on by late capitalist development, perhaps a new definition of the machine is needed to "break down the iron wall" between nature and technology by constructing a transversal relation between them. What Guattari is attempting to do here is nothing less than replace the philosophical concept of techne, which Martin Heidegger appropriated from the Greeks, with the more abstract and encompassing concept of the machine. "The problem of *techne* would now only be a subsidiary part of a much wider machine problematic. Since the machine is opened out towards its machinic environment and maintains all sorts of relationships with social constituents and individual subjectivities, the concept of technological machine should therefore be broadened to that of *machinic assemblages*."[39] The concept of the machine points to a logic of the continuous deterritorialization of elements at the service of particular functions and relations of alterity. Importantly, a machine can readily connect to different orders of being by cutting across the dualities between nature and artifice, object and subject. Guattari claims that capitalism functions on the basis of axiomatization and, more generally, of capturing much more recalcitrant machinic enunciations. This is why he clearly prioritizes affective and presignifying modes of thought to one "which claims to give a scientific, axiomatic description" of things.[40] Lazzarato also argues that "capitalism is neither a mode of production nor a system" but rather "a series of devices for machinic enslavement" that operate by "mobilizing and modulating pre-individual, pre-cognitive, and pre-verbal components of subjectivity, forcing affects, percepts, and unindividuated sensations . . . to function like the cogs and components in a machine."[41] These ideas are beautifully expressed in a work of video art called *Assemblages*, cocreated by Angela Melitopoulos and Maurizio Lazzarato in 2010.[42]

With this, it is easy to understand the importance of the machine and asignifying semiotics for Lazzarato, especially in the era of what he calls immaterial labor. Indeed, the plane of machinic enunciation seems to be the primary field on which a critical contemporary battle is being waged: art against empire. Lazzarato broadly conflates the categories of signifying and asignifying semiotics with Deleuze's differentiation between the respective logics of disciplinary and control societies. He does this by reading these logics through Guattari's idea that capitalism does not operate simply on the economic register, since it is more fundamentally a semantic operator that informs all levels of production and power. On the one hand, signifying semiotics operate through everyday discourse, representation, and the production of meaning in order to give rise to the speaking subject by implicating it into the molar categories of identity, gender, nationality, and class. Guattari calls this process social subjection, and, Lazzarato argues, it also roughly corresponds to Foucault's disciplinary "concept of government by individualization." On the other hand, asignifying semiotics operate through machinic enslavement, a much more insidious, molecular process that captures and activates the pre- and transsubjective elements of percepts and affects to force them to "function like components or cogs in the semiotic machine of capital."[43] This asignifying, molecular level should be understood as being inhabited by prediscursive rhythms, intensities, colors, and sounds that shape the very conditions of image, word, and therefore subjectivity itself. As such, following William James, Lazzarato calls it an unruly world of pure experience, which is precisely the source of its power. Indeed Guattari refers to the elements of asignifying semiotics as "power signs."[44] These signs are understood as material particles that do not pass through linguistic chains, but rather plug into the body directly through preconscious affects, perceptions, desires, and emotions. They do not produce signification or speak but function machinically through "a direct, unmediated impact on the real," which triggers "an action, a reaction, a behavior, an attitude, a posture."[45]

Lazzarato argues that the importance of asignifying semiotics for the analysis of contemporary capitalism cannot be overemphasised. Although it is ignored by most linguistic and political theories, it constitutes the linchpin of new forms of capitalist governmentality. While traditional political theories tend to merely employ logocentric conceptions of enunciation, in post-Fordism we find a growing number of signs that are produced and circulated by machinic apparatuses such as television, cinema, video, and the Internet. Lazzarato's main thesis in *Videophilosophy* is that video art grants us access to the ontology—or more precisely, makes visible the onto-aesthetics of asignifying semiotics—inherent to "the new nature of capitalism."[46] His wager is that we can therefore utilize this technology to somehow help us escape the clutches of contemporary control society and develop new "practices of freedom and processes of individual and collective subjectivation."[47]

But Lazzarato clearly states that art can no longer be deployed, as it was for the historical avant-garde, as a surrogate for forms of political action no longer capable of mobilizing beliefs and desires. This is why he follows Guattari in speaking of an "ethico-aesthetic paradigm" rather than fetishizing the objects and institutions of art. It is also why he avoids the term "politics of aesthetics." Artistic practices are taken as technologies of the self that may be more adequate models for action in our time of immaterial labor, in which the manipulation of subjectivity—understood by Guattari as traversed most fundamentally by non-discursive and affective flows—has become a crucial form of control. In the face of this, ethico-aesthetics asks the question, what forms of *aisthesis* and *poiesis* can most powerfully respond to the seeming indifference of postmodern capital? It is interesting to note that Antonio Negri has developed a similar idea of an ethico-aesthetic paradigm in his book *Art and Multitude*, in which the model of art is understood as an aid in constructing practices that are a necessary first step in conceiving a political program. "The passage to ethics, and therefore the true *potenza* of constructing a meaningful world—this is the way out of the postmodern . . . out of the machine of the market."[48] Neither Negri nor Lazzarato view this task with any kind of romantic nostalgia or utopian hope. Machines that crystallize time may simply help us to more clearly perceive the power-signs at work in the time of post-Fordism. And as machines, they may further help us to creatively reconfigure these signs in order to construct new, and perhaps "monstrous," modes of being in the world—beyond the "existential nullity"[49] of our contemporary landscape—by plugging into other assemblages and apparatuses: "Art is not cut out to play a hegemonic role, but its forces could be joined together with other dispositives (economic, social, political) and other techniques both for the worse (creating a market, becoming a tourist, consumer and communicator of subjectivity and thus contributing to its uniformization) or for the better in making the bifurcations and experimentation of subjectivity possible."[50]

INTRODUCTION

Problem: Where are the barbarians of the twentieth century? Obviously, they will become visible and consolidate themselves only after tremendous socialist crises.
—Friedrich Nietzsche

0.1

This book sets out to analyze the functions of electronic and digital technologies as crystallizations, or syntheses, of time. These functions will be described by staging an imaginary dialogue between the temporal ontology of Henri Bergson and the texts of video artist Nam June Paik in order to show that video technology imitates not nature, but time. The theme of time and its syntheses allows me to, among other things, establish a connection to the Marxian theory of value, in which labor equals time and the commodity equals its crystallization. However, there is an immediate displacement, since machines that crystallize time modulate and capture the time of life and not simply the time of labor. In the time of life, following a Benjaminian distinction, labor and perception tend to become reversible on a new plane of immanence traced by these technological assemblages. The time of life is synonymous with the complexity of semiotics, forces, and affects involved in the production of subjectivity and the world that contemporary capitalism, in its highly sophisticated organizational forms, wants to put to work by capturing their creativity.

0.2

Bergson says that both the ancients and moderns saw in philosophy a "substitution of the concept for the percept."[1] In the tradition of Western thought the substitution of the concept for the image is grounded upon a belief in the inadequacy of our perceptual faculties, an inadequacy testified by the other human faculties; for example, the faculty of understanding and its functions of abstraction and generalization. According to Bergson, the philosophies of the ancient Greeks focused on perception, "since it was by the transformations of a sensible element like water, air or fire, that they completed the immediate sensation" (Bergson, "The Perception of Change," 109). But as soon as philosophy thought it had demonstrated the impossibility of a method of transformation that would remain too close to sensory data, it embarked, according to Bergson, upon a path it would never give up, which leads us to the supersensible world. It is with pure ideas that philosophy henceforth attempted to explain things. But Bergson proposes the opposite path: "Suppose that we were to insert our will into it, and that this will, expanding, were to expand our vision of things" (111). This return to perception—its expansion and extension—is justified, first of all, by the fact that the function of the understanding is to subtract from the real a large number of qualitative differences and therefore to partly diminish our perception. Second, if the intellect is so efficient at separating, choosing, combining, and coordinating, it does so at the expense of the power of creation. Thus, for Bergson, *there is more reality and creativity in the percept and the image than in the concept and the understanding.* This immersion into perception will lead us to a point where nonsemiotically formed matter is reached: time-matter and image-matter. Crucially, what the understanding excludes from perception is not simply qualitative difference; it essentially excludes time. In order for the intellect and the concept to operate, they need to conceal becoming; that is, they need to impose a chronological temporality, which amounts to a reduction of time to space. For the understanding, change and movement can only be understood as the succession of immobile units. This plunge into perception reaches, beyond the intellect, time in the making, detaching itself from the already made and enjoining the act of making.

This new conception of the relationship between time and image, developed through a critique of the privilege that the metaphysical tradition held for the supersensible world, leads to the affirmation of a consciousness and thought of duration, or nonchronological temporality. That is, it leads to an experience from which all the semiotics of subjectivity—perception, image, intellectual labor, and so on—will be defined, beyond the separation between body and mind, as syntheses of time. Can we "ask the eyes of the body, or those of the mind, to see more than they see?" (111). Can we subtract ourselves from language and the errors into which it leads? In response to these questions, Bergson proposes the method of intuition that allows us to move beyond the various oppositions of Western philosophy and access duration. This mental operation, by becoming attuned to its own thoughts, invents machinic apparatuses whose production of images exceeds the constraints of natural perception. In fact, it intersects the relationships between image, time, and the "work of the mind," whose framework has been drawn out in Bergson's *Matter and Memory*.

This book begins with the following hypothesis: these machines, like the mind, crystallize time. And they do so by imitating the two ways in which consciousness arranges the syntheses of time: by its own intensive movement and by the material through which it passes. But this new Bergsonian conceptualization is inspired by a strange "mechanical mysticism" that, while referring directly to the first Industrial Revolution, in fact calls for a second. According to Bergson, the mind and spirit have their source in matter. During the Industrial Revolution, matter developed an "artificial body" that required "another spirit," but this spirit could not develop without "a powerful tool providing the fulcrum." Bergson was convinced that the extended body of the first Industrial Revolution required a kind of supplementary soul. While he grasped the conditions of the development of mechanical and thermodynamic machines, he could not anticipate that the additional soul, the one he deemed necessary, would also take on a machinic form.[2] According to him, the development of the artificial body that "endowed us with powers beside which those of our body barely count" had reached such a level that the "material barrier had almost fallen."[3] Although I cannot follow the path that led Bergson toward a mechanical mysticism and a new spirituality, I will retain and freely use the concepts produced by his quite original passage through the artificial body of the Industrial Revolution.

Finally, I would like to emphasize the power of Bergson's positive ontology for a definition of the society of the image. According to him, for a mind that purely and simply follows the thread of experience there is no void, no nothingness, no possible negation. A negative ontology, on the contrary, can only be the product of a logical philosophy that construes being through nothingness and the negative, since the logical essence of being is precisely what neutralizes time and prevents its apprehension as duration.[4] Based on this positive ontology, the image escapes from lack, absence, and the negative. *It ceases to supplement reality by representing it in order to become, on the contrary, the very fabric of being.*

0.3

The body, which lies at the center of Friedrich Nietzsche's philosophy, is a source of astonishment greater than either consciousness or language. The trajectory of this idea passes through a critique of the splitting between the sensible and the supersensible and thus leads us from the human body toward a superior body, one that incorporates all developments of the mind. Nietzsche firmly rejects the notions that thought is a "thinking of thought" and that we cannot think except through language. According to Nietzsche, most of our movements have nothing in common with an all-too-human consciousness, because if this were not so, they would already be selected, ordered, and semiotized. That which accedes to such consciousness is the result of a conflict between forces that are already marked by utility and value. Thoughts and feelings are extremely "small and fleeting" compared to the innumerable factors that fill even the smallest of instants.[5] Through language, rational thought obscures both its own genealogy as well as any idea that does not comply with its imperatives. What Nietzsche finds in the multiplicity of nonrational forms and nondiscursive thought are the foci and vectors of subjectivation and creativity: preorganic thoughts—"the realization of forms, as in the case of crystals"—consisting of sounds, images, and tactile sensations (Nietzsche, *The Will to Power*, 273). Against the absolutism of the concept, the

Nietzschean line inscribes not only a genealogical reconstruction of conceptual thought but also a questioning of the modes of assemblage and conflict it maintains with other forms of thought. The body then ceases to be construed as an obstacle for accessing the power of thinking. On the contrary, we must plunge into the body in order to grasp the *unthought* within thought—that by which thought exists.

The genealogy of the concept, according to a sequence of forces and signs, reconstructs a continuity between the movements of matter, life, perception, and concept. The separation of spirit from matter results in practices of power that mark the body physically. Similar to how Nietzsche, in some celebrated pages, compared the acquisition of memory to torture that engraves upon the body the memory of a debt or the formation of a "seeing eye," the fixation of perception demanded by the origin of organic life is an unheard-of cruelty, which eliminates everything that is felt differently (275).

According to Nietzsche, thought and perception are symptoms of the will to power. Therefore, vision is not primarily conditioned by the physiology of the eye, but by a will to power that is a force of creation. We are all artists in the sense that there is no natural perception and that "divination" is the method through which perception is organized. This point of view allows us to criticize most models of perception and cognition of phenomenological or rationalist origin. If for Bergson there is more reality in the image than in the concept, then Nietzsche radicalizes this view even further by claiming we must look for this reality within the body and that all those things we have come to know as mind or spirit are in fact grounded in the body. The development of this spirit—as well as, for us, the so-called intellectual or cognitive technologies—is then nothing but a symptom of a superior body ready to be born:

Perhaps the entire evolution of the spirit is a question of the body; it is the history of the development of a higher body that emerges into our sensibility. The organic is rising to yet higher levels. Our lust for knowledge of nature is a means through which the body desires to perfect itself. Or rather: hundreds of thousands of experiments are made to change the nourishment, the mode of living and of dwelling of the body; consciousness and evaluations in the body, all kinds of pleasure and displeasure, are signs of these changes and experiments. In the long run, it is not a question of man at all: he is to be overcome. (358)

And this surplus of reality present within the body, within the social body, within the superior body of which Nietzsche speaks, is in fact capitalist accumulation. It is both the matter and the subject of contemporary capitalist accumulation. The fixed capital of this new form of accumulation is precisely that which captures, semiotizes, integrates, and leverages this surplus reality. New digital technologies do not deprive us of a body, but rather of the illusions of the mind and its powers. They do not impede our vision but demand more power and strength: an ethics of vision. They do not simulate reality but remind us that we constantly utilize things that don't really exist (points, lines, signs, concepts, names) in order to act.

0.4

Most theories of new technology face two major limitations that this book seeks to overcome. The first consists in the relation of exteriority that such theories establish between technology and capitalism. The second is the fact that most of these theories consciously thematize, or implicitly presuppose, an ontology based on the split between reality and concept, between the world and its image. These theories therefore remain dependent upon a theory of representation or a phenomenology that places the object within space and the image within the mind. Contrary to this, I will try to demonstrate how the capitalist machine, including its processes of deterritorialization as well as its constituent and captive powers, lies at the foundation of the genesis and operation of machines that produce images. Indeed, any definition of machines that crystallize time is based on the assumption that time is subordinated to capitalist valorization. Paradoxically, this subordination cannot occur without the liberation of time, since its qualitative aspect is the source of the "continuous creation of unforeseeable novelty."[6] Indeed, the organization of time abstracts economic value; it frees time from all burdens of the natural and the subjective. I am referring here to the Aristotelian crisis of the conception of time, in which time is understood as the measurement of the extensive movements of cosmological nature, and to the Neoplatonic crisis, in which time is conceived as the measurement of intensive movements of the soul, as well as to the emergence of the "whatever-time" specific to capitalism. Derivative time, the time of everyday banality, the time of instants that pass away into nothingness, has captured original time, the rhythm of nature or the movement of the soul, a movement that measured and redeemed. Capitalism has now reversed the subordination of derivative time to original time, such that time no longer measures the harmony of the cosmos or the agreement of the soul. There is only the time of everyday banality, which can no longer be measured or redeemed by any model. But the whatever-time of capitalism takes the form of a specific duplication: whatever-time as the producer of value (Karl Marx) and whatever-time as the producer of continuous unforeseeable novelties (Bergson).

Time in capitalism therefore no longer refers to itself but is founded exclusively as whatever-time by surpassing a certain degree of deterritorialization. Similarly, it is only beyond a certain level of deterritorialization that matter loses its solidity and becomes molecular. The deterritorialization of matter and time gives rise to molecular matter and time as continuous sources of creation, which have always been actualized, but which capitalism alone manages to reveal and organize. Deterritorialized matter and time are the working material for machines that crystallize time, whose functions are determined by the reproduction of the syntheses of whatever-time. From this point of view the engagement with Marx is crucial, since labor and commodities are forms of the capture and crystallization of time. But while Marx indicated the methodology with which to discover "living labor" beyond work, he is of no help in analyzing the forces that lie beneath representation, memory, and perception unless we fall back onto the old dogma of structure and superstructure, which is utterly useless under the conditions of contemporary capitalism.

Turning now to the second limitation, we find that the definition of machines that crystallize time presupposes an ontology that puts into question the splitting of the world into reality and concept, essence and phenomenon, being and appearance, which constitutes the foundation of the modern philosophical tradition. Even Marxian theory and its critique of political economy operate with this splitting of reality (use value) from the concept (value), as well as their reversal. But contemporary capitalism is the realization of this reversal because value has triumphed over, or subordinated, use value. The traditional categories of modern philosophy and the concepts for the critique of political economy have proved to be either powerless or tautological. Bergson and Nietzsche, albeit with some remarkable differences, have constructed theories of perception, affect, memory, and concept that share a desire to oppose all such metaphysical splits between reality and its appearance. In the same way that Marx, and with him the entire revolutionary tradition, studied the European philosophical canon as conceptual personae of the foundation of the ontology of capital—capital as practically existing idealism—Bergson and Nietzsche should be understood as the conceptual personae who have constructed an ontology for contemporary capitalism.[7]

According to this hypothesis on the new nature of capitalism, the technologies of time are not technologies that double reality in representation but apparatuses that operate on a new plane of immanence, a new ontology: the triumph of the appearance of the complete subordination of use-value to value, which liberates time. From the once assumed impossibility of a dialectic between concept and reality, it is necessary to move toward a new perspective opened by Nietzsche: "We find no conflict between reality and appearance, but would willingly talk about degrees of being and, more willingly still, about degrees of appearance."[8]

On this new plane of immanence, according to Bergson, the distinctions made by the philosophical tradition between the objective and the subjective, the extended and the unextended, and quality and quantity cease to be operative since they are all conceived from the point of view of space. Contrary to the postmodern interpretation, the problem lies not in the fact that all difference is rendered impossible or futile, but rather in the fact that it is upon the basis of time that such distinctions must be made. And they must be made not upon the basis of time as a fourth dimension of space but upon a qualitative time, a time that "is happening and, even more, that causes everything to happen."[9] For Bergson, we must still determine the differences, but only on the basis of a relation between the actual and the virtual that grounds time and subjectivity. The machines that crystallize time work within this temporality: they do not represent the world; they crystallize through the relaxation and contraction of time and thereby contribute to its constitution. But to enter this new relationship between body and spirit—for example, how the mind affects the body and vice versa—we must abandon all theories of representation that split being into reality and image as well as any phenomenology that places objects within space and images within the mind.

0.5

The incredible technological development of machines that crystallize time, their powers of deterritorialization and simulation, tends to focus on human faculties like memory, perception, understanding, and imagination. But if these machines imitate human faculties, they do so according to processes other than those described by psychology. Following the Bergsonian genealogy of the capacity for seeing, I put forth the hypothesis that technologies of the image do not reproduce the function of the eye, but rather the subpersonal (pure perception) and suprapersonal (ontological memory) conditions from which images, memory, and understanding are constructed. That is, information and communication technologies do not primarily exteriorize the faculties of the subject but, much more profoundly, the conditions governing its constitution. In the final analysis they are not, as Marshall McLuhan suggested, the prostheses of sense or of the organs and faculties, but products of the will to power and the capacity to act of a body that moves in and by way of time.

The passage from video technology to technologies that simulate thought and action seems to traverse the close connection Bergson and Nietzsche established between body, image, and thought. One video artist describes the phylogenesis of these technologies: "With the integration of images and video within computer logic, we are approaching the ability of mapping the conceptual structures of the brain according to a technological perspective.... After the first video cameras and recorders *provided us with an eye connected to a form of memory that was "coarse" and non-selective*, we find ourselves in the next step of evolution: the field of perception and artificial, yet intelligent, structures of thought."[10] The idea stated here is that of a trajectory which connects models of the eye and ear with models of the processes of thought and ultimately with the functioning of the brain. By relying upon the work of Bergson, I will try to illustrate how perception, sensation, and memory are all products of the capacity to contract and dilate time and that subjectivity is essentially constituted within the actual-virtual circuit. This inhuman production of the human, as the synthesis of time, is also the basis of electronic and digital technologies.

The proof of this assertion will be based upon the works of pioneering video artists like Nam June Paik, who stated without hesitation that video is time: "Video has components of both space and time, but time is now the most important component because the so-called static image is nothing other than lines and because with the electronic image, in fact, there is no space and everything is time."[11] Thus, according to Paik, video imitates not nature but time. This understanding of the imitation of time by the video machine is the ground upon which I will attempt to establish a dialogue with Bergson, since the time in question here is a synthesis of a time released from its reference to the movements of the cosmos and the soul. However, the imitation of time cannot be understood simply as an imitation of the syntheses of time. The path opened up by Paik—in which video must be understood not as the impression of light on a support, but as a synthesis-modulation and contraction-expansion of light—should also be applied to digital technologies. From the production of the cinematic image to the synthetic image, I will therefore draw other connections and indeed ruptures with respect to the theoretical tradition that views the transition from analog (impression of light on a support) to digital (production of the image by a language) as a paradigm shift. One final consideration: the passage from the eye to processes of thought must be reconstructed according to indications that can be found in the methodologies of Nietzsche and Bergson, where the emphasis is not on physiology of the eye, nor on the brain or memory, but rather on the power to act and the affective forces that define it.

0.6

What must be understood by the crystallization of time? We can grasp it as a creative process or as an accumulation of force, since according to Bergson, we should consider duration as "a force in its own way" and time as a "cause of gain or loss."[12] In the domain of life, duration seems to act as a cause, or a quasi cause, because the reversibility of time indicated by classical science can never be realized. But what force does duration exert? It is neither a kinetic nor a potential energy. Bergson subscribes to the materialist tradition according to which force is closely related to sensation. Nietzsche, who based his entire philosophy on the concept of force, defines the will to power as a "pathos" and as a "primitive affective force" from which all other forces derive.[13] Subsequently, Bergson introduced in the qualification of affective force an essential determination: it can only be understood with respect to time as duration. He therefore discovered the relation between affective force, subjectivity, and duration.

Duration acts in two different ways with respect to affective force. First, all sensations take place in time, which is not indifferent to their intensity: "A sensation, by the mere fact of being prolonged, is altered to the point of becoming unbearable. The same does not here remain the same, but is reinforced and swollen by the whole of its past."[14] But more profoundly, duration is the original source of all sensation. Duration, from this point of view, is affect; it is the very capacity to affect and be affected. Indeed, affections and sensations must logically presuppose a deeper force that makes them possible, since "at the most basic and seemingly indivisible level of coloration there is already duration and succession, a multiplicity of conjoining points and instants whose integration is an enigma. By what force are successive sonic moments, where one stops and another begins, combined together? What makes possible this productive coupling of death and life?"[15]

All sensation, developing over time, requires a force that conserves what no longer is. Otherwise it would be reducible to a simple excitation. We must therefore conceive of a force that does not function according to the sensory-motor schema, but according to a duration that retains death within life, the before within the after. Bergson defines duration by this capacity of memory, this capacity for conservation: "It is memory, but not personal memory, external to what it retains, distinct from a past whose preservation it assures; it is a memory within change itself, a memory that prolongs the before into the after, keeping them from being mere snapshots and appearing and disappearing in a present ceaselessly reborn."[16] Without this conserving force of duration, without this productive succession that contracts death in life, the before in the after, there would be no sensation, no time, no accumulation, and therefore no growth. Without this duration, the world would have to begin again in each instant. It would consist of a present that repeats itself indefinitely, but without difference.

The radical novelty that Bergson introduced to the idea of force is this quality of the power of conservation, and therefore of sensation, that traverses the circuit of the actual and the virtual. The relationship between actual and virtual, the ground and motor of Bergsonian duration, determines force as both receptivity and spontaneity. If the actual-virtual circuit lies at the foundation of force in terms of receptivity—since it allows for conservation and memory—it is also the foundation of force in terms of spontaneity, because the virtual always tends toward actualization. This vital force is simultaneously élan, impulse, and memory prolonging the before into the after. The power to affect and be affected has a temporal basis. Duration is therefore a concrete reality. For Bergson time is no longer a way of being, but being itself. He derives an important consequence for the definition of machines that crystallize time: the relationship between the actual and the virtual is a polarized one that releases and produces a specific energy that, following Gilles Deleuze, might be defined as inorganic. "No one can tell whether the study of physiological phenomena in general, and of nervous phenomena in particular, will not reveal to us, besides the *vis viva* or kinetic energy of which Leibniz spoke, and the potential energy which was a later and necessary adjunct, some *new kind of energy* which may differ from the other two by rebelling against calculation."[17] Machines that crystallize time function on the basis of this energy, and their power is derived from the capacity to capture and produce. We must therefore differentiate between crystallizing machines and mechanical or thermodynamic machines in terms of the capacity of both to produce and accumulate affective force and to be grafted onto, in a very specific way, by the processes of the production of subjectivity. Bergson allows us to analyze the relationship between subjectivity and time on the basis of desire and the energy it generates.

0.7

We must now take another step foreword and focus on Félix Guattari's work on modelization in order to access the dimension of contemporary capitalism and its machinic assemblages. Guattari has reinvented the intuitions and concepts of Nietzsche and Bergson by *incorporating the production of subjectivity with the modes of social production and the requirements of valorization in capitalist society*. Of particular interest, with respect to Guattari, is that the matter of subjectivation discovered by Nietzsche and Bergson not only constitutes the vectors and centers of subjectivation through language and concepts but is formed by collective assemblages, which simultaneously operate on the levels of enunciation and the machinic. That is, the apparatuses of the production of subjectivity include the subpersonal and suprapersonal conditions defined by Bergson and Nietzsche as well as machinic apparatuses, collective assemblages, and the social order of capitalism.[18]

Furthermore, in order to start reconstructing collective assemblages for the production of subjectivity in the age of postmodern capitalism, I will draw freely from the suggestion made by Walter Benjamin that cinematic production plays a decisive role in the liquidation of the difference between manual and intellectual labor: "The division of labor came into being at the moment when a difference between manual and intellectual labor appeared. If this affirmation of German ideology is correct, nothing could better facilitate the liquidation of labor in general and the polytechnical education of humanity than the flattening of the separation between manual labor and intellectual labor. This flattening is particularly visible now—if not exclusively, then at least with a heightened clarity—in cinematic production."[19]

In the information economy this flattening has been fully realized. For Benjamin, the form of the reception of the cinematic spectator anticipates the transformation of subjectivity that the information economy captures. On the one hand, cinema introduces a mode of collective perception that "appropriates perceptions of the psychotic and the dreamer."[20] The resulting aberrant forms of movements and images initiate us to an optical unconscious, which renders "the representation of the human by means of an apparatus . . . highly productive."[21] On the other hand, the emotional and spectacular pleasures of cinema arouse "an immediate, intimate . . . attitude of expert appraisal" that profoundly transforms the "modes of participation" of the masses.[22] Benjamin links these transformations in the attitudes of the masses, in relation to collective perception, with the transformations in labor. In doing so, he identifies a reversibility between producer and consumer, author and audience, that now seems to underlie the information economy. According to Marx, his own most important scientific discovery was the concept of labor power, or living labor. This concept allowed him to define a gap, a specific difference, within the determination of the subject upon which he could base both productive force and an ethics. If he had followed this line of thought, he would have completed the project of political economy, as he himself admits. My task here is to analyze precisely this gap, this creative force, in relation to language, communication, information, and representation. Since language, communication, and information are the categories with which the production of subjectivity is controlled and overcoded, desubjectifying them will allow us to critique the ideology and practices of postmodern capitalism.

1

THE WAR MACHINE OF THE KINO-EYE AND THE KINOKI AGAINST THE SPECTACLE

> Kino-eye as cinema analysis,
> kino-eye as the "theory of intervals,"
> kino-eye as the theory of relativity on the screen . . .
> kino-eye is understood as "that which the eye doesn't see,"
> as the microscope and telescope of time,
> as the negative of time,
> as the possibility of seeing without limits and distances,
> as the remote control of movie cameras,
> as tele-eye,
> as x-ray eye . . .
> kino-eye as the possibility of making the invisible visible . . .
> for the communist decoding of the world.
>
> —Dziga Vertov, "The Birth of Kino-Eye"

1.1

Within the work of the revolutionary Kinoki movement in Russia, headed by Dziga Vertov, we are able to witness a functioning war machine posed against the capitalist machine's production of images and representation. Both in theory and in practice the Kinoki anticipate, and proactively shift, the critique of the spectacle as it was conceived in the late 1960s and, in many ways, as it presents itself today. As Vertov attempted to show, to resist the spectacle we must address the machine that produces it. But the notion of the spectacle should not be understood only in the situationist manner, as the receding of the real into representation or as the reversal of the world image. Indeed, the mere recognition of the alienation of the directly lived within the world of images and representation does not in itself relieve us of this predicament. Paradoxically, even within the domain of cinematic technologies (including video and digital media), the Foucauldian genealogy of disciplinary techniques—in which we have neither stage, spectacle, nor representation, but machine—remains valid. And this machine is at once aesthetic, semiotic, technological, and social.

Launching an "assault against the visible world," as it has been organized by capitalism, demands the integration of these different assemblages.[1] Throughout the remainder of the twentieth century, the complexity and radicality—in the sense of getting to the root of things—of this position would be diluted by aesthetic, political, and social criticisms that were largely opposed to and incompatible with each other. Even Jean-Luc Godard's Dziga Vertov Group, which referred explicitly to this experiment, turned out to be nothing more than a pale and timid allusion to the complexity of the Vertovian vision.

1.2

Vertov interpreted the Soviet revolution not only as the general collapse of power and of capitalist institutions in Russia but also as a collapse of the human and its world. Cinema was immediately understood as a machinic expression of forces—assembled in different ways with human forces such as seeing, feeling, perceiving, and thinking—capable of opening up new forms of subjectivation. These forces of the outside that capitalism brings to the foreground are, in the first place, forces of time and of the virtual that can be located within machines that crystallize time as powerful means of expression. The camera liberates perception and thought as well as the center of gravity that defines the human body. The *kino-eye*, which Vertov also called the machine-eye, moves in a perpetual metamorphosis—a discontinuous movement of bodies—rendering sensible new matter, new affects, and new forces. Thus, in the intensity of the first cinematic images, the world is shaken and seems to lose its solidity and stability. In this becoming of bodies, the kino-eye captures their intensity, their incorporeal element, introducing not only movement but also time into the image. Aberrant movements of the camera, along with montage, introduce us to a direct experience of time, to durations and speeds that are no longer merely human.

By grasping, in the becoming of bodies, the virtuality of a deterritorialized world, cinema alludes to a new body and a new thought. With cinema we have become mutants of perception, vision, and thought in the same way that within factories the human has become irreversibly hybridized with mechanical and thermodynamic machinery. We discover with horror or with joy that we do not think with an all-too-human consciousness, but rather through machines: seeing machines and thinking machines. At the beginning of the cinematic adventure, this visual thinking renews Gottfried Leibniz's and Baruch Spinoza's attempts to grasp the spiritual automaton that connects images within and beyond the human. The Kinoki movement refers to this as the terrain of class struggle.

1.3

Capitalism brings forth a new visibility in which the subject of the "I see" is not a psychological subject and whose social form is not reducible to the public. We have instead a collective and multiple subject that cannot be understood simply as a frequent customer of dark cinema halls. Yet Vertov's new proletariat is not just an ideological reference point but primarily a new aesthetic and productive paradigm of assemblage: the factory instead of the theater, the cyborg of the collective worker, and the assemblage of human and machine, instead of purely subjective representation. Deciphering the visible cannot be undertaken by literary, theatrical, or graphic technologies, but only by a machine, the cinematic machine. The "I see" of the kino-eye involves a process of singularization of the collective body of the proletariat, which simultaneously arranges forces within the human, technological apparatuses, temporal flows, and social assemblages.

For Vertov, the first requirement for experiencing this transformation in terms of class is to ensure that cinema does not merely remain enclosed within itself but rather grasps the temporal specificity of these new machines and their immediately social assemblages. Normally, the filmmaker, the producer, and the audience collaborate more or less consciously in reproducing their own specific roles, thus reinforcing the functions of subjugation and enslavement characteristic of the capitalist machine of image production and representation. Similarly, the mass character of cinematic communication is a formal condition of cinematic communication itself, which must be integrated as such in film production.

1.4

Vertov denounced the idea of enclosing the cinema within itself as an operation in which commercial and artistic forms seize power at the expense of other possible forms of production. "Existing cinema, as a commercial affair, like cinema as a sphere of art, has nothing in common with our work" (Vertov, "The Birth of Kino-Eye," 74). The idea is not to produce different content—whether artistic, political, or social—within the cinematic structure but to destroy cinema as a capitalist machine for the production of the sensible, of perception and thought, on its own terrain. "Long live class vision!" does not refer to a vision of the world that is more ethical, political, or aesthetic but rather to another type of corporeal and technological assemblage of enunciation (66). Only then can all cinematic functions be reassembled, resulting in a complete change of their nature.

Vertov saw in a particularly acute manner that the machinic and social processes of creation, the virtualities of new forms of perception and thought, which worldwide class struggle inspired, are constantly folded back into the spectator-director relationship. The film drama—with its actors, screenplays, studios, and directors—is the form in which representation captures new means of expression and diminishes the powers of the collective body that global revolution has brought into the public sphere. Vertov knew neither what this new body could do nor what collective forms of expression this newly constituted mutant subject, the industrial proletariat, was capable of. But he did know that cinesensibility was a major political issue.

1.5

All the polemics between Vertov and Hollywood, as well as his more nuanced controversy with Sergei Eisenstein, were organized around the need for the revolution to subtract cinema from its images and representation. Vertov wanted to explode the technological assemblage and the division of labor in cinema, because both fold the forces of time with the forces within the human back into mere representation and image.

Cinema, with its static shots and standardized rhythms, risks merely gratifying our eyes that "see very poorly and very little," instead of exploring the chaos of visual phenomena that fill the space, irrespective of the position of our body during observation (67). "I am the kino-eye. I am a mechanical eye. I, a machine, show you the world as only I can see it. Now and forever, I release myself from human immobility, I am in constant motion. . . . Freed from the rule of sixteen to seventeen frames per second, free of the limits of time and space. I put together any given points in the universe, no matter where I've recorded them."[2] The Kinoki program develops the accidents of filming—rapid shot, microshot, mobile shot, shots that employ the most unexpected angles, and so on—into a system of apparent aberrations that push us toward the perception of time. "The kino-eye is defined as . . . the microscope and telescope of time."[3] The kino-eye leads to a new perception of the world; it decodes with fresh eyes an unknown world.

Within the cinematic division of labor, a screenplay functions as a normative apparatus that removes any evental dimension within cinematic practice. "Kinopravda doesn't order life to proceed according to a writer's screenplay, but observes and records life as it is and only then draws conclusions from these observations."[4] The screenplay, Vertov insists, is the invention of a single person or group of people, not the evental encounter with a world that is unknown to us. "Proceeding from material to film object, and not from film object to material, the Kinoki seize the last (most tenacious) stronghold of artistic cinema in the literary screenplay."[5] The simulation of an event by a director is of secondary importance compared to the inactuality of the real time of life. These two perspectives require completely different qualities and organizations of labor.

If the camera is the machinic eye that allows us to be in continuous movement, to flow into the perpetual variation of bodies, montage should not reduce this new visibility to human perception and its prejudices (the psychology of the eye and its linguistic fetishism, as Friedrich Nietzsche would say). Montage aims at the "organization of the visible world" that respects the temporal dimension and the forces that constitute it.[6] "Attempts in this direction have been made. And it must be said, with some success. Editing tables containing definite calculations, similar to systems of musical notation, as well as studies in rhythm, 'intervals,' etc."[7]

In the projection room—"Stupefied by the opium of bourgeois film dramas,"[8] "intoxicated by cine-nicotine, the spectator sticks like a leech to the screen that tickles his nerves"[9]—the Kinoki prefer factories, trains—"I'm in charge of a cine-train. We're showing films in a remote station"[10]—and ships. In an evocation of Franz Kafka, they combine means of communication that allow us to travel in time—images or words—with those that allow us to travel in space.

1.6

Vertov tirelessly repeats that the purpose of the kino-eye is to see, to make see—"Kino-eye opens the eyes, clears the vision"[11]—because we are not yet seeing. The kino-eye allows us to relate a movement and an image at any point in the universe, through an appropriate reaction, to a different movement and image at any other point in the universe. These images and movements are incommensurable and imperceptible from the point of view of the human eye. "Kino-eye is the possibility of seeing life processes in any temporal order or at any speed inaccessible to the human eye."[12]

The kino-eye must discover in the chaos of movement the "resultant force,"[13] as yet unknown, and must produce from "all these mutual interactions, these mutual attractions and repulsions of shots," from the whole "multitude of 'intervals' . . . a simple visual equation, a visual formula expressing the basic theme of the film object in the best way."[14] The objective is to engage directly with the study of the visual phenomena that surround us and that constitute life. "'Art and everyday life' interests us less than the topic, say, of 'everyday life and its organization.'"[15] Vertov was acutely aware that what cinema announces is the reorganization of the processes of the production of subjectivities around machines that crystallize time.

The realization of this project requires a social organization and technology that do not reproduce the usual division of labor in cinematic production. In this regard, Vertov projected a production in six series. All the realized cineworks of the Kinoki are concerned only with the first series, which he called "Life Caught Unawares":[16]

> In this part the camera, having chosen some easily vulnerable point, cautiously enters into life and takes its bearings in its visual surroundings. In subsequent parts, along with an increase in the number of cameras, the area under observation will be extended. Gradually, through comparison of various parts of the globe,

various bits of life, the visible world is being explored. Each succeeding part will further clarify the understanding of reality. . . . Millions of workers, having recovered their sight, are beginning to doubt the necessity of supporting the bourgeois structure of the world.[17]

The same visual material would pass progressively to deeper analyses and reorganizations that highlight the relations of subjects by using all the technical and formal means available to the cinema. For Vertov, the "factory of facts"[18] requires cineobservers who produce cineobservations and cineanalyses within the framework of a poetic cinema. Film ultimately abandoned this possible becoming, and it is only in the work of video artists that we can find a reinvention of this methodology.

Cinesensation is interpreted by Vertov as a constitutive force. Cinelinkage, connecting the proletariat of all nations through audio-visual means (kino-eye), is an anticipation of television, digital, and satellite networks, which power and the markets utilize for production and enslavement. "Kino-eye means the conquest of space, the visual linkage of people throughout the entire world based on the continuous exchange of visible fact, of cine-documents as opposed to the exchange of cine-theatrical representations."[19]

1.7

The originality of Vertov's critique of the image as reification of the visible lies in the fact that, for him, the genetic element of the visible is the interval. The visible consists not only of images but also what lies between them: jumps, cuts, rhythms, and other aberrant movements. "The school of kino-eye calls for construction of the film-object upon 'intervals,' that is, upon the movement between images."[20] "Intervals (the transitions from one movement to another), are the material, the elements of the art of movement, and by no means the movements themselves."[21]

The interval (jump, cut, fade) is thus a connection that functions to suture, cover, obliterate, or tame not our all-too-human eye but the nonimaged ground, the deterritorialized flux from which images emerge. The interval, irreducible to the image and movement, is on the contrary their source, their origin, and their cause. The interval is that which is not reducible to the discursive and the figurative. We can hear Bergsonian accents in this effort to exceed representation and images through the interval: "The primary material of the art of movement is by no means movement itself, but the intervals, the transitions, the transitions from one movement to another."[22] The kino-eye is a machine for the contraction-relaxation, or crystallization, of time: "The mechanical eye, the camera . . . gropes its way through the chaos of visual events, letting itself be drawn or repelled by movement, probing, as it goes, the path of its own movement. It experiments, distending time, dissecting movement, or, in contrary fashion, absorbing time within itself."[23] If, as the situationists have claimed, "the whole life of society . . . presents itself [as] an immense accumulation of spectacles," and if "the spectacle is capital accumulated to the point where it becomes image," we must push our analysis beyond the commodity-image.[24] Karl Marx identifies, in the relationship between time and subjectivity, the key that uncovers the enigma of labor and the commodity as the crystallization of time. With Vertov's films and writings, I propose to reflect upon a different crystallization of time. Cinema is the first of a new type of machine that fixes and reproduces the time of perception, sensibility, and thought.

1.8

Cinema shows in practice that thought overflows human consciousness in the same way that images overflow perception. Humans have lost the confidence in being the sole producers of thoughts and images. Therefore, what is at stake in the era of the collapse of the human and its world is the power of thought, or the image of thought, and its processes of creation.

The visual thinking that the kino-eye realized aligns itself with the spiritual automaton that, as in Bergson, provokes circuits of ideas within memory and opens the possibility of breaking down thoughts directly—without passing through linguistic semiotics—upon the screen in the spectator's brain. "Thoughts fly from the screen to viewer, without having to translate thought into words. It is a living relationship with the screen, a transmission from brain to brain. . . . Each penetrates into a circuit of ideas that stirs the seeds of its own consciousness."[25] Cinelanguage, which Vertov opposed to spoken and literary language, refers to the complexity of forces and signs that work to produce thought.

> The point is that the exposition of *Three Songs* [*About Lenin*] develops not through the channel of words, but through other channels, through the interaction of sound and image, through the combination of many channels. It proceeds underground, sometimes casting a dozen words onto the surface . . . develops in a spiral fashion, now in the sound, now in the image, now in a voice, now in an intertitle . . . now through movement within the shot, now in the collision of one group shot with another, now smoothly, now by jolts from dark to light, from slow to fast, from the tired to the vigorous, now through noise.[26]

This image of words thrown onto the surface shows that cinelanguage is a powerful tool for the repudiation of the imperialism of signifying semiotics, which imposes a fetishism of subject and object onto the production of thought. Words written and spoken in a film have a path traced in counterpoint in relation to asignifying semiotics, an entirely different image of thought that the kino-eye projects on the screen.

1.9

The nonhuman perception of the kino-eye points toward a new human, a kind of Übermensch that in Vertov's view has nothing to do with communist humanism. For Vertov there is no opposition between human and machine, because he assumes a second nature produced by capitalism, both as an irreversible reality and as a condition by which to move beyond the human. Kino-eye and radio-ear—today it is perhaps more accurate to speak of a computer brain—are hybrids with which the collective subject of the revolution must see, speak, and hear; a machinic body, a cyborg of vision, perception, and thought that must express itself as such, without mediation. To the technological and financial concentration of cinema, the kino-eye responds with a micropolitics that implies the socialization of know-how as well as a miniaturization of technology.

> We have absolutely no need of huge studios or massive sets, just as we have no need for "mighty" film directors, "great" actors, and "amazing," photogenic women. On the other hand, we must have:
>
> 1) quick means of transport
> 2) more sensitive film

3) small, lightweight, hand-held cameras
4) lighting equipment that is equally lightweight
5) a staff of lightning-fast film reporters
6) an army of kinok-observers

In our organization, we distinguish amongst:

1) kinok-observers
2) kinok-cameramen
3) kinok-constructors
4) kinok-editors (women and men)
5) kinok laboratory assistants

We teach our methods of cinema work only to Komsomols and Young Pioneers; we pass on our skill and our technical experience to the rising generation of young workers in whom we place our trust.[27]

Vertov believed that the mass character of cinema should not be limited to its distribution alone but should also emphatically include its production to insure that its power of expression is not expropriated. The critique of cinedrama found its complement in the practical critique of the concentration and control of the means of production by the film industry. From this point of view, Soviet power merely reproduced the organization of labor it wanted to criticize. Against the cinema of the left and its commitment, Vertov proposed a micropolitics that could provide Soviet workers the possibility of constructing a strategic cinelinkage for the processes of collective subjectivation. The collective form of cinema had its demands: "The departure from authorship by one person or a group of persons to mass authorship will, in our view, accelerate the destruction of bourgeois, artistic cinema and its attributes: the poser-actor, the fairy-tale script, those costly toys and sets, and the director-high priest."[28]

1.10

Vertov was perhaps alone in thinking of and organizing the cinema not as an art of the masses but as a mass activity, a constitutive activity. He didn't work as an artist, but as a relay within a network of correspondents scattered throughout the Soviet Union. He was working within a flow that spilled over on all sides, that he could not and did not want to control. This conception of labor problematizes the division between intellectual and manual labor and, as a consequence, the conception of the artist, author, and intellectual. "This film represents an assault on our reality by the camera and prepares the theme of creative labor against a background of class contradictions and of everyday life."[29]

The work of the Kinoki cannot be considered artistic labor but should be recognized within the larger production of being and the collective subject. "The red cell Kinoki should be regarded as one of the factories in which the raw material supplied by kinok-observers is . . . drawn into the production of future cine-works."[30] Of course, Vertov's position has nothing to do with the anti-intellectual and populist ideology of the proletarian artist (the proletarian director as a counterpart of the proletarian writer). What he does claim is that the kino-eye contains assemblages that open unknown territories beyond the author and artist, becomings that contain virtualities for other aesthetic, social, and productive paradigms.

1.11

The method of the war machine consists in the act of pushing further the deterritorialization of flows, the human-machine hybridization, the concentration and decomposition of time, and the development of the cinelinkages, in order to reterritorialize them onto a new body. If we push the cinematic assemblage toward a higher level of deterritorialization and socialization, we have video. And indeed the assemblage of cinema technology is utilized by Vertov as an anticipation of video. From the birth of the revolutionary experiment of the Kinoki, Vertov's films organize and think with and through the teledistribution of images and sounds: "From the human eye's viewpoint I haven't really the right to 'edit in' myself beside those who are seated in this hall, for instance. Yet in kino-eye space, I can edit myself not only sitting here beside you, but in various parts of the globe. it would be absurd to create obstacles such as walls and distance for kino-eye. In anticipation of television it should be clear that such 'vision-at-a-distance' is possible in montage."[31] For Vertov, television is not only a technological assemblage more adequate for the deterritorialization of flows, their circulation, their cutting and splicing, but also a technological assemblage more adequate with regard to the social and collective dimension of the production of life, which capitalism has introduced as one of its preconditions. "The method of radio-broadcasting images, just recently invented, can bring us still closer to our cherished goal ... the establishing of a class bond that is visual (kino-eye) and auditory (radio-ear) between the proletarians of all nations and all lands, based on the platform of the communist decoding of the world."[32] Vertov is not concerned with any of the problems that arise for world cinephilia in relation to television, since for him it is a completely natural and necessary development of the productive, perceptive, and creative forms of cooperation of the collective body that capitalism and the class struggle evoke. When posed in these terms—though television did not yet exist as a technological apparatus—the problem of the telebroadcasting of images and sounds avoids all the false debates between art and popular culture.

1.12

The concepts of kino-eye and radio-ear owe nothing to the futurist fascination with machines, to which they are often reduced. Instead, they are precise articulations of both the new conditions for the production of subjectivity and the relationships to technology that this new configuration determines. But for Vertov, the social machine always takes precedence over the technological machine. "Even in technique we only partially overlap with so-called artistic cinema, since the goals we have set for ourselves require a different technical approach."[33] It is the particular and specific montage of social and technological assemblages, aesthetic forms, and networks that demands certain technologies and not the other way around. The work of the Kinoki anticipates and necessitates new technological assemblages. "The theoretical and practical work of the Kinoki (in contrast to acted cinematography) was in advance of our technical possibilities; they have long awaited the overdue (in relation to kino-eye) technical base for sound film and television."[34] In the technologically backward situation of the Soviet Union in the midtwenties, the Kinoks anticipated television.

1.13

The impossibility of developing this micropolitical war machine in the revolution and Hollywood's crushing of any becoming of cinema with the cinedrama, whether commercial or artistic, had its logical consequence in the films of Leni Riefenstahl. Soviet power preferred to have artists like Eisenstein rather than a war machine, which implies a radical questioning of the concept of power. The formidable potential that the revolution's forms of communication contained was developed into the spectacle. Although the situationists were able to identify this new form of domination, they barely scratched the surface of the conceptualization of the war machine, which was necessary to oppose it. Every time cinema wanted to call itself into question (neorealism, the New Wave), it had to tap into this virtuality of the kino-eye. And with each new development of technological apparatuses that capture and reproduce, in an increasingly sophisticated way, the relationship between time and subjectivity—the real time of the event and the virtual—the questions left open by the failures of the Vertovian vision had to be posed again.

1.14

The Kinoki movement attempted to assemble machines to see, feel, and think within a process of constructing a collective body beyond propaganda and artistic representation. Barred from the outset, this experiment did not tell us what this new body could do. But now that new technological apparatuses, new machines that crystallize time, are once again proposing a superior body, the kino-eye can always remind us of what we must resist:

- The collapsing of deterritorialized flows and forces of time into representational images.
- The collapsing of new processes for the production of subjectivity into public spectatorship.
- The collapsing of new forms of cooperation and new forms of knowing, feeling, and thinking into figures of the author or artist.

Our response to the questions left open by Vertov's experiment becomes even more urgent in our time, in which creative labor effectively assumes a social and universal dimension beyond the separation between intellectual and manual labor, and in which perception and labor are realized with the same machines. But Vertov always reminds us what debates about new technologies too easily avoid: new technological assemblages demand the creation of new collective assemblages of enunciation. In order to safeguard all the promises that they periodically announce, these machines for seeing and thinking must remain open to all other semiotics and to all other forms of subjectivity and temporality that the multitude and cosmos express.

2

BERGSON AND MACHINES THAT CRYSTALLIZE TIME

2.1 BERGSON AND DURATION TIME

2.1.1

I have chosen to focus on the work of Henri Bergson because it gives us a description of natural perception as a relationship between flows of images, between durations and different rhythms. This relationship between flows is functionally guaranteed by the body, consciousness, and memory, which operate as genuine interfaces by introducing a time of indetermination and elaboration as well as a choice into the stream of flows. But video and computer technologies function according to the same principle: into the stream of flows they insert a cut, an interval that allows for a specifically machinic organization of the relationship between signifying and asignifying flows. The functional relation is guaranteed here by a technological assemblage.

In Bergson, this relationship between flows is grounded in the capacity to act as well as in affective force. He thus allows us to rid ourselves of the problematic of the disappearance of the real and the visible, since he assures us that they have always been a function of our capacity to act. He allows us to pose the only reasonable question that can be addressed to these new technologies: to what degree of power, to what capacity to act, do they correspond?

Finally, Bergson shows us the multiplicity and heterogeneity of the elements and assemblages that constitute subjectivity and the higher activities of the mind—the solidarity and conflict between the movements of matter, consciousness, and the body on the basis of time. By thinking differently about the genealogy of the faculties and the configuration of forces, he allows us to construct a critique of the concepts of subjectivity and intellectual labor. But we will begin with Bergson and video. Video is, in fact, the first technology corresponding to a generalized decoding of the flows of images, whose genesis can be traced in Bergson's *Matter and Memory*.

2.1.2

Bergson develops conceptual tools to apprehend the specificity of the production of images in the capitalist era; namely, that images are automatically produced by technologies and are thus withdrawn from human activity. The production of images not only becomes automatic and therefore independent of humans but also introduces movement and duration as its ontology. Capitalism and its technologies introduce movement and time into images, and vice versa.[1] This theme can also be found throughout the work of Walter Benjamin. But in Bergson, the image-movement-time relationship allows us to move beyond the ambiguity of Benjamin's concept of technological reproducibility, which wavers between the industrial mass reproduction of the unique work of art and the exploration of the regimes of temporality proper to capitalism.

All of Bergson's work is a sustained reflection on time and its force: "I perceived one fine day that time served no purpose, did nothing. Nevertheless, I said to myself, time is something. Therefore it acts. What can it be doing? Plain common sense answered: time is what hinders everything from being given at once. It retards, or rather it is retardation. It must, therefore, be elaboration. Would it not then be a vehicle of creation and of choice? Would not the existence of time prove that there is indetermination in things? Would not time be that indetermination itself?"[2] This completely reverses the theoretical positions (for example, Paul Virilio's) for which the real time of electronic and digital technologies prevents any delay, gap, waiting, or indeterminacy. The technologies of time, by increasing the possibility of retaining and conserving time, should develop our power and capacity to act. If power and action seem, on the contrary, neutralized, we will have to look for the reasons somewhere other than in the technologies themselves. The time evoked by these theories inherits the idea of spatialized time, conceived as the fourth dimension of space and not as duration; that is, nonchronological time. I will use Bergson's work here to provide a definition of time as the continuous creation of unpredictable novelties.

2.2 HABIT AND MEMORY

2.2.1

If Nam Jun Paik's assertion that video imitates not nature but time is true, then the technologies I am interested in should imitate various contraction-syntheses of time. The operation of contraction-synthesis allows for the conservation, accumulation, and production of time in two different forms: habit and memory.

We should first distinguish, in Deleuzian terms, a material synthesis that contracts successive, independent, and actual instants or elements from a spiritual synthesis that contracts all levels of a coexisting, virtual past. The first operates according to discontinuous material vibrations (light and heat being Bergson's examples) by constituting discontinuous durations and producing perception-images. The second operates according to time (the virtual) by constituting continuous durations and producing memory-images.

Material and spiritual syntheses might thus be called memory. But the first is a habit rather than a memory. It conserves the past in the sensory-motor mechanisms of the body that "play rather than represent it." Only the second can truly be called memory, because it conserves the past in independent recollections that intervene not automatically but intelligently in the constitution of the image and the realization of an action. The material synthesis is therefore a form of automatic or passive recognition, while the spiritual synthesis is a form of active or intelligent recognition.

But for Bergson, intelligent or active recognition is consciousness, and consciousness means, first and foremost, memory.[3] All consciousness is the conservation and accumulation of the past in the present, the retention of the before in the after, the integration of "the dead into the living." I will highlight two aspects of this capacity to conserve and accumulate time. First, memory is the creation of a gap, a temporal interval between movements that are received and movements that are executed by each body. Second, within this gap there arises and develops a force that acts and exploits this delay, this moment of indetermination between action and reaction. This force is simultaneously will and sensation, spontaneity and receptivity, memory and habit.

My reading of Bergson is quite particular, since it focuses on the relationship between force and time. In Bergson consciousness-memory is the capacity to act and the capacity to feel, two inseparable yet distinct aspects of the same force. "Immanent in the inward life, it feels rather than sees it, but feels it as a movement, as a continual treading on a future that recoils without ceasing."[4] Bergsonian duration is a force and acts like a force because it produces the capacity to feel, to be affected. The first function of perception is precisely to grasp a series of elementary changes in the form of a quality or a simple state by the work of condensation or synthesis. "In the smallest discernible fraction of a second, in the almost instantaneous perception of a sensible quality, there may be trillions of oscillations which repeat themselves. The permanence of a sensible quality consists in this repetition of movements."[5]

There is a direct relationship between the capacity to act and the capacity to synthesize, which, it should be emphasized, is not exclusively an anthropomorphic force: "The greater the power of acting bestowed upon an animal species, the more numerous, probably, are the elementary changes that its faculty of perceiving concentrates into one of its instants. And the progress must be continuous, in nature, from the beings that vibrate almost in unison with the oscillations of the ether, up to those that embrace trillions of these oscillations in the shortest of their simple perceptions."[6] But every memory is also the anticipation of the future and of action. If on the one hand, memory is the capacity to conserve, to accumulate, then on the other it operates as a force with the capacity to affect, to produce images, and to act. The conservation and accumulation of the past takes place according not to what is but to what will be. The power to act is always an encroachment on the future. To retain what no longer is and to anticipate what is not yet: these are the functions of consciousness. In Bergson's work the before and the after, the past and the present, should be interpreted not only as successive determinations of time but also as conditions of power, of affective force in the form of activity and passivity. Thus memory produces an energy whose nature must be sought in an extraspatial process, since it is an affective energy, a powerful, nonorganic energy, as Gilles Deleuze describes it.

The material and spiritual syntheses are thus the metamorphoses of this original force that is immediately qualified by duration, both in its aspect as receptivity—to conserve the past within the present—and in its aspect as spontaneity—to tend toward the future, to act and create. According to Bergson, power, both active and passive, is duration. Machines that crystallize time, by retaining and accumulating duration, can help either to develop or to neutralize the force of feeling and the force of acting; that is, contribute to our becoming active or becoming passive.

2.2.2

On the concept of material synthesis, Bergson distinguishes natural perception from pure perception. Natural perception is not privileged and should never be the starting point for apprehending the world. According to Bergson, our psychological representation starts by being impersonal and always maintains this ground of exteriority. Everything becomes clear if we begin with exteriority and move toward the interior, whereas problems multiply if we begin with the center—the body and the subject—and move toward the periphery.

It is therefore necessary to abolish consciousness and move beyond the human turn of experience that stretches a homogeneous space and time over that which exists. The abolition of consciousness leads to the abolition of our perception being organized according to the necessity of action.

> The material universe subsists exactly as it was; only, since you have removed that particular rhythm of duration which was the condition of my action upon things, these things draw back into themselves, mark as many moments in their own existence as science distinguishes in it; and sensible qualities, without vanishing, are spread and diluted in an incomparably more divided duration. Matter thus resolves itself into numberless vibrations, all linked together in uninterrupted continuity, all bound up with each other, and traveling in every direction like shivers through an immense body.[7]

The passage beyond the human turn of experience moves us into pure perception; that is, the identity of matter, image, and movement. The world is nothing but a vortex of images. There is only the flow of images that encounter, collide, reflect, compose, and decompose one another. In this dimension of forces, intensities, and becomings—"Condense atoms into centers of force, dissolve them into vortices revolving in a continuous fluid"[8]—the world is a flow of light. The world as image should be understood precisely in the sense of streaming flows of light, of an infinite variation of pure vibrations.

In this coexistence of all images without center, direction, or orientation, we should understand that the image does not take the form we usually attribute to it. This is the image in itself, an image that no eye can perceive. In fact, we are dealing with images and perceptions of things themselves.[9] The image, in pure perception, is nothing but a center of action that receives and transmits movements in which action and reaction merge. All images act and react upon each other. And this action-reaction is accomplished, not by an element that is specialized for this function (the eye, for example), but by all the elementary parts of the image at once. The image is defined by shock, pure vibration, shivers. This is obviously a metaphor, since pure perception only exists de jure, and as Bergson suggests, this vision of matter is perhaps too "fatiguing" for our imagination and remains hidden from our understanding. We can perceive contractions of pure perception, but never pure perception as such.

But it must be emphasized that matter is identical not only to the image, to the perpetual variation of images, but also to time.[10] Pure perception is thus defined by a series of equivalences: image = movement = light = matter = time. From this series of equivalences, there will emerge remarkable points or bodies that, by the intervals that define them, constitute sensation and perception through material and spiritual syntheses.

Material synthesis operates on this matter, which is a continuum of time-images, by producing perception: "Perception seizes upon the infinitely repeated shocks of light or heat, for example, and contracts them into relatively invariable sensations: trillions of external vibrations are what the vision of a color condenses in our eyes in a fraction of a second."[11] Memory, or spiritual, synthesis is the capacity to bind together these independent moments, to embed them into each other and constitute them as duration, as rhythm. Through this operation, consciousness both makes contact with matter and distinguishes itself from it. If this duration is abolished, the result will be an infinitely more divided and diluted duration. The image of pure perception is an infinitely relaxed or dilated duration.

According to Bergson, the subject and object, as well as their forms of perception and representation, are compositions (contractions) of flows of images—defined as pure shock punctuating a present that eternally begins again—that overflow on all sides. Subject, object, and perception do not produce images but are, instead, contained within the flows of images.

Bergson's work appeals directly to our "society of the image" since, on the basis of time-matter and image-matter, it offers a model of the constitution of subjectivity and the world. Image-matter is not presented here as a signifying or representing element, but as the true genetic element of the world, which implies a real twist in relation to all other analyses of the split between the real and the representation, between the sensible and the intelligible. The novelty of this ontology lies in the fact that it never reduces perception to a duplication of the real, since perception is not in us but in things. Being and appearing coincide, but appearing is an appearing in itself, which does not require a perceiving subject.

I cannot help but think that the matter processed by video machines is close to something like pure perception. In fact, a concept of perception as the contraction of "infinitely repeated shocks of light or heat" corresponds precisely, as we shall see, to what video technology does.

2.2.3

Perception arises from pure perception and image-matter. In the perpetual and acentered variation of all the images that constitute pure perception, remarkable points are formed that, by reflecting images, select a portion of them and thus determine gaps or intervals in the continuum of time-matter. These remarkable points are special images: bodies. Bodies, as images, simply receive movements, select them, and transform them into action, because the gap in which bodies are constituted introduces a time, an indetermination, a possibility of choosing between action and reaction. The receptivity and spontaneity produced in this interval are no longer, as in pure perception, identical. The human body is distinguished from other bodies only by the more complex and elaborate form through which it actively responds to a received action. "My body is, then, in the aggregate of the material world, an image which acts like other images, receiving and giving back movement, with, perhaps, this difference only, that my body appears to choose, within certain limits, the manner in which it shall restore what it receives."[12] In pure perception, therefore, perception does not move from one body to another but, since it is first disseminated throughout the ensemble of bodies, gradually limits itself and adopts the body as center. And this action-reaction is achieved, as the center becomes more and more complex, by a specialized part of the body; for example, the eye and the brain.

In a leap that crosses millions of years, Bergson moves from the gap of the first infinitesimal body that is constituted in the prebiological soup—whose pure perception can be regarded as a form of reconstitution—to the most advanced and complex expression of the gap or interval, namely the brain.

The brain, as a part of the body, functions only as an interface, but an interface through which the interval between received and performed movement is maximized. Actions no longer immediately follow from a received action but have the possibility of being based on a selection (receptivity) of the latter and developing a delayed action (spontaneity of new actions) on the basis of this delay, this time gained by the development of the brain and nervous system. The more complex this system, the greater its capacity to act.

Significantly, Bergson defines the brain, using the technological metaphors of his time, as a "telephone exchange office." Its role is to establish communication or put it on hold. In other words, for Bergson the brain operates as an instrument of analysis with regard to received movement and as an instrument of selection with regard to performed movement. The brain, therefore, does nothing but continue or transform the flows of light. It is contained within universal variation and does not create images, does not add to the perception of things. Quite the contrary, its function is to subtract and retain from image-matter that which is useful for its needs, necessary for its action. Our images are therefore not something we add to the object, but rather a selection, or subtraction, from matter. Image and perception are a slice of the continuous flow, an arrest of movement. In short, the brain is not a creative center of perception and conscious representation.

Images are therefore not produced by the brain. If the world is a flow of images, if perception is in things, then the "cerebral state" is within images and perception, and not the reverse. "It is neither its cause, nor its effect, nor in any sense its duplicate: it merely continues it, the perception being our virtual action and the cerebral state our action already begun" (Bergson, *Matter and Memory*, 232–33). The brain is just an interface in the sense that it translates one speed into another, one movement into another; an interface that translates the infinite flow according to the needs of an action. It is a relay between different degrees of the real.

The video camera is also a body that has been plunged into the time-image, that creates its own interval within it—capturing and crystallizing the perpetually varying flows of pure perception—and that constructs a more or less delayed action with respect to received actions.

2.2.4

Consistent with his plane of organization of the world, Bergson should determine the constitution of sensation starting from the pure perception within the gap conditioned by the body: "Affection *must*, at a given moment, arise out of the image" (55).

Bergson reverses the priority of sensation over perception. Whereas by reducing perception to a weakened sensation, psychology makes the material universe an association of our subjective states, since perception would then be nothing more than an exteriorization of the internal states of individual consciousness, Bergson proceeds differently. The body, itself an image, is first of all a center of perception and movement, a center of perception of other images, and a center of action-reaction to the stimuli of other images. According to Bergson, within the interval that constitutes the body, between perception and reaction, a third moment that constitutes the affective state is inserted. As I have indicated before, these operations of perception and action-reaction are produced by all parts of the body as image. The amoeba provides a good example of this situation: "Every part of the protoplasmic mass is equally able to receive a stimulation and to react against it; perception and movement being here blended in a single property—contractility" (55).

But as organisms gradually evolve, parts of bodies, within the division of labor specific to them, abandon action in response to stimulus and conserve "a kind of motor tendency in a sensory nerve" (55–56). The sensitive element is the relative immobility of a very long succession of elementary vibrations on a nervous plate that, instead of reacting immediately, absorbs them, thereby introducing between perception and reaction a sensation; that is, an indetermination, a delay, a possibility of choice. Sensation and affective force are the transformation of extensive movement into the intensive movement brought about by the body. "The moving body has lost its movement of extension, and movement has become movement of expression. It is this combination of a reflecting, immobile unity and of intensive expressive movements which constitutes the affect."[13]

The Bergsonian explanation of the transition from a perception that occupies extension to an affection that is considered unextended is as follows: perception measures our possible action upon things and, inversely, the possible action of things upon us. This action—for example, a danger—is defined as the virtual action of things upon us. But as the distance between a perceived object and our body decreases, the virtual action tends to be transformed into real action: "But the more distance decreases between this object and our body (the more, in other words, the danger becomes urgent or the promise immediate), the more does virtual action tend to pass into real action. Suppose the distance reduced to zero, that is to say that the object to be perceived coincides with our body, that is to say again, that our body is the object to be perceived. Then it is no longer virtual action, but real action, that this specialized perception will express, and this is exactly what affection is."[14] The capacity to receive, integrate, and transmit movements, thereby producing sensations, belongs to all of life. The difference is only a difference in the capacity of a body to act: "The greater the body's power of action . . . the wider is the field that perception embraces."[15] The temporality of bodies consists of a combined system of sensations and movements. This present is, in essence, sensory-motor.

The fact that affection arises from the time-image is a crucial point, since this is the only way to explain how affective force, or Nietzschean pathos, and the indetermination of the power to act are constituted.

2.2.5

Why, for us, does pure perception exist de jure rather than de facto? Because we only understand it as a limit case. In reality, perception is always already memory. Bergson says that we only perceive the past. Since our body is not a mathematical point, but always a duration, it is necessary to introduce memory and its capacity to insert the past into the sensory-motor mechanism, which functions only in the present.[16] Duration is not an ineffable subjective experience, but a precise and determined function of our capacity to perceive and imagine: "The duration lived by our consciousness is a duration with its own determined rhythm, a duration very different from the time of the physicist, which can store up, in a given interval, as great a number of phenomena as we please."[17] To insert time means to accumulate, conserve, and introduce it into the present in order to create an indetermination, a delay. This accumulation-conservation of time is what allows us to act.

Bergson distinguishes two types of memory. The first—automatic or passive recognition—is fixed within the body as habit rather than as real memory. Here, the past is conserved in the motor mechanisms of our organism. Strictly speaking, this memory is without image and confines itself to transforming received movements into preformed movements. The second memory—attentive or intellectual recognition—is the real memory in which the past survives within independent recollections. It should be noted right away that, unlike the first, this memory is not installed within the body but exists in time.

If, as we have seen, images are not produced by the brain, they are also not stored by it. Memory—unlike the brain, which functions as an interface between movements—functions as the interface between the virtual and the actual. Memory no longer continues movement; it continues duration. Memory begins as automatic recognition; that is, through movements, through excitations of time-matter. But while in habit, movements prolong our perception for the purposes of action—and therefore distance us from the perceived object—in attentive perception they bring us back to the object in order to emphasize its contours. Memory, unlike the sensory-motor activity of the body, produces increasingly more precise true images of the perceived object.

How does attentive recognition function? Having determined a gap, we fill in the perceived object by scanning our memory for similar images and experiences. The perceived object, or actual image, triggers a process in which we search for a virtual image and these two images run after each other, thereby determining a new perception. The new perception is constructed through this continuous movement between perception and memory. External perception only arouses movements that trace the broad contours of the image. Memory then directs onto the received perception "memory-images which resemble it and which are already sketched out by the movements themselves. Memory thus creates anew the present perception, or rather it doubles this perception by reflecting upon it either its own image or some other memory-image of the same kind."[18]

It should be emphasized that the process of recognition cannot be reduced to the simple association of a perception with a recollection, but is rather something like a divination or even a hallucination.[19] Associationism reduces perception, image, and idea to stable, completed things that it would suffice to associate with each other. Bergson speaks, on the contrary, of movements, forces, and powers that, as they come into reciprocal contact, produce images and ideas. I will conduct a more thorough analysis of this process in chapter 4.3.3. Attentive recognition operates by "the projection, outside ourselves, of an actively created image, identical with, or similar to, the object on which it comes to mold itself."[20] The work of attentive recognition recreates not only the perceived object but also the increasingly vast system to which it is attached.

Finally, it should be noted that the difference between automatic and attentive recognition is determined by conditions that either free memory or make it captive to the fulfillment of the finalized action. Attentive recognition is only possible if memory ceases to be completely engaged in sensory-motor activity and its purposes. This liberation can only be carried out thanks to the intervention of machines—such as the brain, language, and technologies—that replace memory and intellectual labor in the execution of motor-habit. Consequently, freed from finalized and predetermined work as well as confronted with a range of possibilities, memory can redefine itself as virtuality. Automatic recognition is a prisoner to sensory-motor habits, while attentive recognition faces an indetermination and a choice. As we shall see, machines that crystallize time, with their automatic production of images and durations, increasingly free attentive recognition from the habits incurred by natural perception and allow memory to redefine itself as duration, as time in the making, and thus as force.

2.2.6

Bergson introduces a radical break with regard to the apprehension of perception:

> Attentive perception is often represented as a series of processes which make their way in single file; the object exciting sensations,

the sensations causing ideas to start up before them, each idea setting in motion, one in front of the other, points more and more remote of the intellectual mass. Thus there is supposed to be a rectilinear progress, by which the mind goes further and further from the object, never to return to it. We maintain, on the contrary, that reflective perception is a circuit, in which all the elements, including the perceived object itself, hold each other in a state of mutual tension as in an electric circuit, so that no disturbance starting from the object can stop on its way and remain in the depths of the mind: it must always find its way back to the object from where it proceeds. (Bergson, *Matter and Memory*, 103–4)

Perception is not an impression or a recording—for example, of light on a medium—but a construction in which we participate through an active work of synthesis. Bergson explicitly invites us to understand memory not as a drawer or a register in which we could search for images that correspond to a received stimulus, nor as a recording such as a phonogram, but as a work of synthesis between the body and memory. The work of synthesis, or intellectual labor, produces these specific virtual objects, and all intellectual efforts—recollection, understanding, creation—can be triggered by a real or virtual object.

The theory of the production of images in Bergson is not an optical, but rather a temporal, theory that can be explained by the different forms of contraction-relaxation of time; namely, the different syntheses of intellectual labor. Optical phenomena are reducible to sensory-motor movements that will never be able to account for visual and, more generally, intellectual activities. These two paradigms, optical and temporal, entail two radically different conceptions of intellectual labor. In the first, activity is mechanical; it does not interfere with and does not require a transformation of the system. In the second, "on the contrary, an act of attention implies such a solidarity between the mind and its object, it is a circuit so well closed that we cannot pass to states of higher concentration without creating, whole and entire, so many new circuits which envelop the first and have nothing in common between them but the perceived object" (104).

This must be emphasized because most theories of new technology are constructed on the model of vision, in which, by contrast, the optical construction of the image plays a fundamental role in relation to intellectual activity. Bergson counters the paradigm of the impression of light on a medium with an ontology of the expression of light. In both cases there is light, but in completely different modalities that refer to completely different philosophical traditions. The paradigm of the impression of light ultimately refers to the demiurge who looks at the model and engraves a copy on a wax tablet. The temporal paradigm refers to the process of the synthesis of time in order to explain the production of the image. Syntheses belong to the philosophical tradition that begins with the Neoplatonists, passes through Immanuel Kant, and arrives at Bergson.

In line with Bergson, it seems to me that technologies of vision force us to denaturalize our mode of apprehension. Bergson already suggested, at the turn of the last century, that vision is first of all determined by the power to act of the living and that, therefore, it is always necessary to subordinate the mechanical and automatic aspects of the production of images to power, to relationships between forces, to time. Bergson criticizes the optical model for this reason, because it always stops with sensory-motor mechanisms and never attains power.

> Thus, to return to the example of visual perception, the office of the rods and cones is merely to receive excitations which will be

subsequently elaborated into movements, either accomplished or nascent. No perception can result from this, and nowhere in the nervous system are there conscious centers, but perception arises from the same cause which has brought into being the chain of nervous elements, with the organs which sustain them and with life in general. It expresses and measures the power of action in the living being, the indetermination of the movement or of the action which will follow the receipt of the stimulus. (64)

The error of the optical model of the apprehension of vision is that it subtracts the eye and the retinal image from a continuous and integral process that involves "the brain, nerves, retina, and the object itself," which, we should recall, are all images for Bergson. "By what right, then, do we isolate this image to sum up in it the whole of perception?" (215). We do not see with rods and cones. We see with memory and intellectual effort, which are forms of the accumulation and conservation of time. Vision is constructed, developed, and enhanced within temporal syntheses.

2.2.7

Let us now continue the reconstruction of the Bergsonian perspective. While in automatic recognition we always remain on the same plane—that of the present—memory allows us to pass through different planes. We continuously pass from the past (the virtual) to the present (the actual) and vice versa, the present being simply the most contracted moment of our past. This work of memory, which Bergson defines as intellectual labor, consists in the contraction-relaxation of time, of the time of memory. When we perceive, we contract our past to bring it into contact with our perception, and when we remember, we relax it in order to install ourselves within its different levels. It is with regard to this attentive memory that we can understand both functions of memory: contraction-memory and recollection-memory. Recollection-memory conserves every detail of our life by transforming it into a recollection; contraction-memory renders possible perception and the image by contracting and relaxing the layer of recollection that duplicates our life.

We will have to return to this aspect of the activity of the mind—defined by Bergson as intellectual labor, or the labor of spiritual synthesis—at greater length because of its dual interest: it provides an intensive description of both attention and intellectual effort. And it will ultimately allow me to put forth the hypothesis that the power of technologies lies in the fact that they reproduce this activity of the contraction-relaxation of time and therefore reproduce intellectual labor.

2.3 ONTOLOGICAL MEMORY: PURE OR VIRTUAL

2.3.1

Attentive memory is therefore composed of two different forms, contraction-memory and recollection-memory, but it presupposes the existence of a nonpsychological memory, an ontological memory. Bergson calls this pure or virtual memory to distinguish it from psychological memory. Psychological memory needs an ontological memory that splits it and without which the remembrance, conservation, passage, and emergence of time would be impossible.

Bergson defines pure or virtual memory as the coexistence of the actual and the virtual, the present and the past. But the simultaneity of present and past within pure memory occurs in two significantly different forms. Time is the object of a double grounding within ontological memory: it is grounded, on the one hand, in the primacy of the past—in which the present is conserved in the past—and on the other hand, in the primacy of the present, in which the past is the recollection of the present. This twofold grounding of time enables us to account for the paradoxes of the Bergsonian theory of temporality: the past is conserved within itself, and time splits in each moment into a pure present and a pure past.

These dimensions of pure memory are of particular interest because, though technologies of time can only imitate ontological memory in a very limited way, they can work upon their temporalities. Indeed, on the one hand, television and digital networks constitute a memory—in which the present is conserved within the past—and on the other hand, by functioning in real time they work upon the splitting of time; they intervene upon time in the making.

The first form of coexistence of the past with the present—or the virtual with the actual—under the primacy of the past, explains how the past is conserved within itself. Indeed, the conservation of time presupposes a pure past, a past that has never been present. For me to say that something has passed, I need a past that is not a former present, since otherwise the former present would no longer be, precisely because it has passed.

To say that something has passed, we need a pure form of the past, an always already there. It is in this sense that the past and the present are contemporaneous. "Our most distant past adheres to our present and constitutes with it a single and identical uninterrupted change."[21] The primacy of the past over other times means that the past holds true for all time, since in this case, the present is the most contracted form of the past.

But the present and the past, the virtual and the actual, coexist in another form that accounts for the passage of time, which causes it to emerge and renew itself continuously as the splitting of time. The coexistence of the present and the recollection of the present are added to the coexistence of all levels of the past and the present (the most contracted form). Here the past, as a recollection of the present, is nothing more than a duplication of the present. The circuit that makes the present adhere to the most distant past must now be replaced with a circuit in which the present is doubled with its own recollection. We remain in the presence of a pure past but one that, in this case, does not represent an always already there but the recollection of the present, its double. Here the present holds true for all time—the past being nothing but the double of the present—a present that tends toward the future, a time that creates rather than conserves.

Bergson explains remembrance by its splitting, at every moment, into a pure past and a pure present. The fundamental statement that accounts for this paradox is the following: the formation of recollection is not subsequent to but simultaneous with perception. Indeed, if the past and present were conceived only according to the before and after, how could the past recollect the present? For the present to be remembered, present and past must be given together, simultaneously. The past must be concomitant with the present, and no temporal interval can separate them. If the past were not contemporaneous with what happens, how could it conserve recollection?

But pure memory is also what allows time to emerge and pass. In fact, to conceive the past as the former present, as that which comes after, implies a spatial and instantaneous definition of the present, unable to access the true nature of the present, which is in the making. Here the present, in order to pass, supposes an ontological memory, a coexistence of past and present. "For the present moment, always going forward, fleeting limit between the immediate past which is now no more and the immediate future which is not yet, would be a mere abstraction were it not the moving mirror which continually reflects perception as a memory."[22] Bergson affirms without hesitation that time implies a succession. But he denies that this succession is constituted by the juxtaposition of a before and after.[23]

The relationship between actual and virtual image (psychological memory) is therefore split by this pure (ontological) memory, which constitutes the fundamental operation of time: its perpetual splitting, its capacity to distinguish and differentiate itself, to be the internal cause of its own differentiation. "Our actual existence, then, whilst it is unrolled in time, duplicates itself all along with a virtual existence, a mirror-image," because our life is not an abstract mathematical point but a duration.[24]

The importance of this concept of the splitting of time merits our staying with it a bit longer. To properly understand the relationship between actual and virtual, we must distinguish the memory-image in the process of actualizing itself from pure memory. Between the two, there is a difference in kind. Pure memory is the memory that does not correspond to any previous experience, being only the duplication of our present.

As I have already indicated, to explain remembrance and the passage of time, we must presuppose a past that "no interval separates from the present" (Bergson, *Mind-Energy*, 138). This pure memory is useless in terms of action—which is why we are unaware of its existence—but is indispensable to the ontological foundation of memory and time. "But what can be more unavailing for our present action than memory of the present? Rather would any other kind of memory be entitled to lay a claim, for it at least brings with it some information, though it be of no actual interest. Alone, memory of the present has nothing to teach us, being only the double of perception. We have the real object, what are we to do with the virtual image of it?" (142). It is this pure memory that Bergson also calls "virtual" and to which he opposes psychological memory.

Each moment of our life thus presents two aspects: actual and virtual, perception on the one hand and recollection on the other. It splits at the same time as it arises. Or rather, it consists in this very splitting. The actual and virtual image are contemporaneous; they are constituted simultaneously but are different in kind. The circuit that the actual object and the virtual image draw as their limit and foundation is the circuit between the actual and the virtual—between the present and the recollection of the present—in which recollection no longer refers to the object that gave rise to it, but only to the present of which it is the double. Bergson uses the image of the actor—an image that Friedrich Nietzsche uses to define the last stage of the will to power—in order to metaphorically account for the creative power of time. "Whoever becomes conscious of the continual duplicating of his present into perception and memory . . . will compare himself to an actor playing his part automatically, listening to himself and beholding himself play. The more deeply he analyses his experience, the more he will split into two personages, one of which moves about on the stage while the other sits and looks" (135).

It is easy to grasp the value of this conclusion for my argument. The splitting of time allows us to establish an absolute difference and thus escape the dualities of the sensible and the intelligible, essence and phenomenon, which can no longer ground difference. This continuous splitting of time is the difference that expresses itself, the otherness that immediately expresses the object to which it is immediately adjacent. To concrete and real time Bergson opposes abstract time conceived as the fourth dimension of space, as pure change, the heterogeneity that produces itself, the nonchronological time of pure memory that reveals the determination of difference.

2.3.2

To conclude this quick reconstruction of Bergson's thought, let us return to the relationship between time and force, between time and the capacity to act, because it is through this force and capacity to act that technologies of time intervene.

The actual-virtual circuit, as the producer of absolute difference, can also be called power. As I have already emphasized with regard to ontological memory, the before and the after, past and present, are not exclusively successive determinations of the course of time but the conditions of power, affective force, and nonorganic energy. Bergsonian duration, and the virtual-actual relationship that grounds it, is the originary affective force itself (Nietzschean pathos). The novelty that Bergson introduces to the definition of affective force is that its heterogeneity is the heterogeneity of time.

The virtual-actual relationship, under the primacy of the past, determines the capacity to integrate, to contract the before into the after, that which is no longer into that which is, the "dead into the living." This memory that conserves is constitutive of all capacity to feel. But the virtual-actual relationship, under the primacy of the present, determines a duration that, instead of conserving, tends toward its actualization. This duration is élan, impulse, tendency, force. "Here again the present is perceived in the future on which it treads, rather than apprehended in itself" (135). Force is qualified as receptivity and spontaneity by the temporal dimension. Deleuze, by interpreting the bifurcation of time as the splitting of power into affecting and affected, defines subjectivity by this temporal dimension, following Bergson. Time as inner sense—being both action and passion—is the affection of the self by the self. Force in all its expressions—passion, action, autoaffection—is characterized by duration.

On the basis of Bergson's research, Deleuze reconstructs three syntheses of time that are simultaneously three forms of subjectivity, since they are three different ways of living in time. The definition of the three Deleuzian syntheses is of paramount importance since it highlights the limits of the concepts of perception and affect, which are usually used to account for the effect of technologies. And even more so because it systematizes the ontology of time as well as its force of organization and constitution of the world.

According to Deleuze, the capacity to perceive presupposes a passive synthesis of time or an originary sensation that constitutes it. Deleuze, directly citing Bergson, distributes the passivity and activity of force in the constitution of the world and subjectivity differently. Perceptive syntheses presuppose organic syntheses, without which there would only be stimuli that bodies could not retain and that would disappear as soon as affection took place. "We are made of contracted water, earth, light and air—not merely prior to the recognition or representation of these, but prior to their being sensed. Every organism, in its receptive and perceptual elements, but also in its viscera, is a sum of contractions, of retentions and expectations."[25]

The perception of an environment indeed presupposes the prior contraction of these elements, even if contraction remains implicit or is concealed by representation and the urgency of action. This primary vital sensibility is already a duration and therefore a force. Every living being is already duration, time. In the cell, for example, the future appears through necessity and the past through cellular heredity. It is only through these passive syntheses of time that active syntheses of memory can be constructed. The simplest operation of the mind, which consists for Bergson in the capacity to bind or contract vibrations and shocks, belongs to all living beings. It is in this sense that we can speak of protosubjectivities.

Already for Bergson, our perception grasps only the surface movements of sensation and matter and ignores the immense multiplicity of movements that it carries out "within itself as a chrysalis. Motionless on the surface, in its very depth it lives and vibrates."[26] Movement is everywhere, especially in the depths, but we only locate it at the surface, which "leads us to distinguish . . . distinct bodies."[27] According to Bergson, if we presuppose an "inert matter" we will never explain how life and consciousness are able to insert themselves and act within it.

> The ancients had imagined a World Soul supposed to assure the continuity of existence of the material universe. Stripping this conception of its mythical element, I should say that the inorganic world is a series of infinitely rapid repetitions or quasi-repetitions which, when totaled, constitute visible and previsible changes. . . . Thus, the living being essentially has duration; it has duration precisely because it is continually elaborating what is new and because there is no elaboration without searching, no searching without groping. Time is this very hesitation, or it is nothing.[28]

Developing the Bergsonian concept of habit, Deleuze argues that this form of duration concerns not only the habits that we have (psychologically) but also the habits that we are. "On the contrary, we have seen that receptivity, understood as a capacity for experiencing affections, was only a consequence, and that the passive self was more profoundly constituted by a synthesis which is itself passive."[29] Thus we are composed of a thousand passive, organic habits. The protosubjectivity of the living is characterized by duration. This first synthesis is called passive synthesis, habit, the present. The passive synthesis constitutes time as present, as the present that passes. But following Bergson, we must presuppose another time in order for the present to pass. This is the ontological time that Deleuze, like Bergson, calls "Memory." "Habit is the originary synthesis of time, which constitutes the life of the passing present; Memory is the fundamental synthesis of time which constitutes the being of the past (that which causes the present to pass)."[30]

But the first two syntheses of time—the synthesis of the present, or "customary cycle," and the synthesis of the past, or "memorial cycle"—require a third synthesis: a time of the future. Deleuze calls this third synthesis of time the empty form of time or, in his later works, the crystal of time. This concept of the crystal of time is derived from Bergson's actual-virtual circuit and Félix Guattari's work on Proustian refrains. Of particular interest for us here are the consequences that Deleuze derives from this concept with respect to subjectivity and affective force. Not only does Deleuze emphasize the possibility of grounding difference with respect to time—the crystal of time being in perpetual self-distinction, a distinction in the making that takes distinct terms into itself in order to endlessly begin again—he also argues that, strictly speaking, the only subjectivity is nonchronological time. "Subjectivity is never ours, it is time.... The actual is always objective, but the virtual is subjective: it was initially the affect, that which we experience in time; then time itself, pure virtuality which divides itself in two as affector and affected, 'the affection of self by self' as definition of time."[31] Affect is time in two different modes. The first corresponds to the contraction of the matter-image on (and by) the body, as Bergson demonstrates. The second is the affection produced by the splitting of time. In conclusion, Deleuze derives from Bergson another definition of power as the actual-virtual circuit.

This conclusion is important for our reflection on the technologies of time. Perception, time, and affect are not products of subjectivity, but on the contrary, it is subjectivity that is internal to perception, time, and affect. Subjectivity is a fold of these impersonal forces. We attach a particular importance to this new ontology because with it, we can easily recognize the condition of our subjectivity in the everyday relation it maintains with the media and technologies of time; namely, that it is we who are interior to time and not the reverse.

2.4 VIDEO AS MATERIAL AND SPIRITUAL SYNTHESIS

2.4.1

After this hasty reconstruction of some Bergsonian concepts, perhaps it is necessary to indicate where this conception of time might lead. I suggest the hypothesis that the technologies of time imitate, in their operations and in their products, the various syntheses of time (the conservation, passage, and splitting-emergence of time) and that, through the operations of contraction-relaxation, they work on the conditions of the production of affective force. The matter that these technologies contract is, as in Bergson, time-matter (material vibrations) and the different temporal stratifications of memory. I use the term *imitation* to signify that electronic and digital technologies operate like the material and spiritual syntheses in Bergson: they crystallize time. Video and digital technologies can therefore be understood as technologies that imitate perception, memory, and intellectual labor.

First, they trace a plane of consistency comprising variations and perpetual perturbations that give us access to something like Bergsonian pure perception. Thereafter they operate simultaneously on a single plane (the present), such as the body, and on several levels (the past), such as the mind. Technologies simulate corporeal perception, since they operate on the single plane of the present like a mechanism that receives and returns movements. Video and digital technologies operate, in the first place, like the Bergsonian brain; namely, like a relay that introduces a gap and transformation into the stream of flows. Like the body and material syntheses, they have the characteristic of returning received movements in a particular way. Indeed, they contract and relax time-matter by transforming an asignifying flow into a signifying flow. They crystallize, by contracting and relaxing, the pure vibrations, shocks, and shivers of matter into images. Moreover, in this operation they are in no way limited by the physiology of the body and can therefore contract and relax beyond the human turn of experience.

Second, machines that crystallize time simulate the labor of the mind and the spiritual syntheses (memory and intellectual effort) by moving across infinite levels of time. Contraction and relaxation no longer concern time-matter or image-matter but the past. If video technology, with its image-processing techniques, reproduces the labor of image production that is unique to memory (albeit in a way that is still quite crude, as Bill Viola suggests), the techniques of simulation seem to be modeled on the description of intellectual labor proposed by Bergson. And finally, through all these syntheses, electronic and digital technologies work, by operating in "real time," on the emergence of time, its conservation within memory, and its unfolding (time in the making). *Video, photography, cinema, and digital technologies are machines that contract and condense time.*

Let us look at these definitions more specifically. Strictly speaking, video technology is not, as it has often been described, a technique for recording light on a medium. The concept of modulation allows us to comprehend video more precisely. The video camera modulates, through apparatuses of contraction and relaxation, the flow of electromagnetic waves. Video images are contractions and relaxations, vibrations and shocks of light, and not tracings or reproductions of reality. The video camera's shot is a crystallization of time-matter that is made possible by technological mechanisms of conventional coding. It is possible to operate a camera without recording and to work on the flows live. It is therefore not at all about the impression of light on a medium but rather about crystallization and its modulation.

The recording, which fixes this modulation onto a medium, allows for the conservation of an image that—being the contraction-relaxation of a flow of vibrations and shocks—can be worked to infinity. The recording does not fix an image, but the vibrations and shocks that constitute it. In video montage we work on light as such and not on images or shots. Montage technologies force us to move inside an artificial memory (even though it is still rudimentary). It is in this sense that this aspect of the video machine refers back to intellectual labor, as the capacity to construct and reconstruct an image to infinity. Technologies for the simulation and production of synthetic images imitate, in a more faithful way, the circuit of intellectual labor, or "synthetic work," and its power of creating and producing time. The relationship between actual and virtual image—with its capacity to construct and reconstruct images and with its infinite proliferation—has been objectified within precise limits in a technological apparatus. Here we no longer have perception-images, but rather memory-images and mental images. It is no longer about constructing an image from the stimuli of raw perception, but simulating the capacity to produce mental images. Of course, what is not reproducible is virtual memory, the actual-virtual circuit as a precondition for the passage of time and the basis for the capacity of recollection. In light of this Bergsonian conceptualization, the definition of these technologies as technologies for crystallizing time takes on its full meaning.

This time, along with its power, is the time of technological machines only because it is, first of all, the time of the social machine of contemporary capitalism. The machines are grafted and constituted on the pure memory—according to the two forms of the conservation and emergence of time—of the social time of capitalism. The actualization of virtuality suggests that it is necessary to place ourselves within this mechanism, where time emerges and is created, in order to master it. Indeed, the actual-virtual circuit can only be controlled by assemblages that operate within this relationship. Mechanical and thermodynamic machines cannot attain this *poietic* core of the production of time. The "live" of television and the real time of digital networks accompany and produce the splitting-emergence of time. The critical importance of this third synthesis of social time consists in the fact that it distributes the past and the future.

Consequently, the entire effort of machines—such as television and also relatively new apparatuses like the Internet—will tend to control, exploit, and channel the power of this temporality. When it is understood that machines of communication function primarily at this level and not only at the level of the signifier, we will have taken a step toward defining a politics of the virtual. Machines that crystallize time are at the heart of the processes of the production of subjectivity, since time, which as we have seen is subjectivity itself, is the power to affect and be affected. Both in terms of memory and intellectual labor, the actual-virtual circuit constitutes the motor of subjectivity.

2.4.2

Before continuing this work on Bergson, I would like to highlight one thing that has been mentioned in passing. Within the framework I have just reconstructed, perception depends upon the power to act and not the reverse.[32] Perception is a function of action; therefore, the limits of perception are the limits of action. This methodology enables us to depsychologize and denaturalize the problem of perception. For Bergson, perception never has to answer to human psychology, which allows him to avoid the methodology, still dominant today, that consists in analyzing perception from within the human-nature paradigm: the perceiving subject and the perceived object (the variations are numerous but always fall within this schema). In this paradigm the world is predefined, and it is only within this given world that perception can take place. Bergson invites us, on the contrary, to understand the forces that, in one and the same movement, give rise to the perceiver and the perceived. The world is not predefined, already given, but is instead constituted by the capacity to act. Perception is nothing but the capacity to increase our power to act, because the effort to increase that power is inseparable from the effort to maximize perception and sensation.[33]

2.5 MACHINES THAT TRIUMPH OVER MECHANISM

2.5.1

I have argued that the machines that crystallize time are the first technologies to remove the human hand from the production of images and render the process automatic. I have also argued that the precondition for this rupture in the history of humanity is represented by the industrial reproduction of time, since images, from any point of view, are time. The development of photography, cinema, video, and the digital is, from this perspective, the development of a motor that, instead of producing and accumulating kinetic and potential energy, produces and accumulates duration, time, and therefore a new kind of energy: affective energy. We know how important the invention of motors has been for the first Industrial Revolution. We can therefore imagine the potential importance of the realization of this quite particular motor that, by becoming independent from will and affective force, either liberates or annuls them.

2.5.2

According to Bergson, the history of humanity and the evolution of nature can be described as the production of a panoply of machines that triumph over mechanism and an automatism that restores the possibility of a choice.[34] The invention-force that mobilizes time, in order to deploy itself, requires a machine that frees consciousness from the completion of teleological action in which it risks being drowned. Bergson proposes to read the development of technological apparatuses in relation to the capacity they develop to "absorb" or "distract" the attention of consciousness in the completion of teleological action.

Bergson is not referring to individual psychological attention, but to the force that senses and acts by producing images and different states of consciousness. Attention is the force and intellectual effort is the conatus of our inner sense that together actualize the interpenetrating virtual images into distinct and juxtaposed images.

On the one hand, absorbing attention in the completion of teleological action means nullifying the gap between what is done and what could be done; it means canceling the interval, the hesitation, and the choice that define consciousness.[35] On the other hand, increasing the possibilities of the "distraction" of attention from sensory-motor activity means increasing the capacity of creation. All machines (concepts, language, technology) have the task of loosening the mechanisms that make us captives of teleological action. Therefore, they operate toward the liberation or neutralization of consciousness and of the possibility of choice.

While Bergson often opposes consciousness and intellect as two antagonistic modes of perception—the first perceives by continuity and becoming, the second perceives by instantaneous cuts and discontinuous states—he is also forced to consider how duration is introduced into our capacity for action by the intellect itself, thereby thematizing its opposition to consciousness differently; because for us consciousness and the real only exist in the tension and oscillation between the virtual and the actual, and not in their separation. The virtual, as an enormous multiplicity of interpenetrating virtualities, can never be known, sensed, or made into the source of our action without its passage to the act; the actual, without its virtualities, would appear as an eternally frozen world with no possibilities for action or change.[36] Society, language, concepts, and technological apparatuses are therefore machines that simultaneously "fix" becoming and allow access to duration, that simultaneously neutralize the actual-virtual circuit in an eternal present and open onto the continuous creation of unforeseeable novelties. This relationship can be usefully reconstructed in the case of language. For Bergson, language is a fetishism—to speak like Nietzsche—that makes us live in a world of illusions, because it freezes and renders becoming static. But language is also a machine that has contributed much to liberating the intellect, thereby contributing significantly to bringing us closer to consciousness. "Without language, intelligence would probably have remained riveted to the material objects which it was interested in considering. It would have lived in a state of somnambulism, outside itself, hypnotized on its own work."[37]

Language liberates the intellect from the completion of teleological action and simultaneously provides consciousness with an immaterial body in which to incarnate itself. It is in this way that language participates in two modes of perception and expression: intellect and consciousness.[38] This liberation of the intellect from sensory-motor habits contracted in action is realized through the mobility of words. It is because the word is mobile, is essentially free and displaceable, that it allows for movement between one thing and another, such that "the intellect was sure to take it, sooner or later, on the wing, while it was not settled on anything, and apply it to an object which is not a thing and which, concealed till then, awaited the coming of the word to pass from darkness to light."[39] A word can move from one perceived thing to another perceived thing, but also to the recollection of it and to the representation of the act by which we represent it; that is, to the idea. According to Bergson, this is how through language the world in itself, and not merely a world as seen from the outside, opens to the intellect. "It profits by the fact that the word is an external thing, which the intelligence can catch hold of and cling to, and at the same time an immaterial thing, by means of which the intelligence can penetrate even to the inwardness of its own work," thereby approaching consciousness and duration.[40]

Now electronic and digital technologies are machines that can deepen our penetration into the interior of duration, beginning with language and the intellect. But unlike words, they have a mobility that concerns images, the elements that compose images—durations and rhythms—and even more profoundly, the syntheses of time.

2.5.3

Electronic and digital technologies, like cinema before them, are machines for the automatic production of the image. Echoing one of Gilbert Simondon's intuitions, rather than defining them as mere externalizations of human senses—like a lens in relation to the eye—it is possible to understand them as motors that have a relative autonomy with respect to the human. Unlike mechanical and thermodynamic motors that take kinetic and potential energy from outside, these motors accumulate and produce duration and time, and therefore affective energy.[41] If memory can be defined as a living motor that accumulates and produces time, video and computers can be defined as technological motors that function according to the same principle.

With these motors that function with affective energy, the image, subtracted from the human hand, acquires a mobility and a relative autonomy, which not only releases memory and imagination from the inert schemas that characterize image production but also increases their capacity for creation. By penetrating into the inner workings of perception, memory, and imagination, electronic and digital technologies—and the forms of knowledge they imply—push us to the limits of cognition and action, as organized by the intellect, and require a paradigm shift that depends upon a different relationship between consciousness and the intellect, between body and mind.

The more intellect advances in its work of analysis, the more it sees the number of heterogeneous elements that constitute the world increase. In the same way, it is led to discover the real continuum that constitutes life. It does this with its method of defining juxtaposed elements, each of which is external to the others. But paradoxically, through this work it tends to go beyond itself and approach real continuity. This is particularly evident in the so-called cognitive sciences. The impasse of research into artificial intelligence demonstrates "how intellectual molds cease to be strictly applicable; and on the other hand, by its own work, it will suggest to us the vague feeling, if nothing more, of what must take the place of intellectual molds. Thus, intuition may bring the intellect to recognize that life does not quite go into the category of the many nor yet into that of the one; that neither mechanical causality nor finality can give a sufficient interpretation of the vital process."[42] But the intellect and its instruments (science and logic), despite the power of their analyses of the heterogeneity and multiplicity of life, and although they have begun to include time and continuity in their methodologies,[43] remain incapable of attaining the real change and radical creation that constitute life, thought, and becoming.[44]

How can we escape this impasse in which the intellect and science push their analyses of physical operations ever further without ever managing to grasp what is most specific to life? At the very moment that the intellect and science go beyond themselves, digital technologies—and especially research in artificial intelligence—encourage us to follow a wish expressed by Bergson: rather than oppose aesthetics and intuition to intellect and science, we should conceive of "an inquiry turned in the same direction as art, which would take life *in general* for its object" (Bergson, *Creative Evolution*, 114). Instead of opposing them, it would be better to "squeeze them both in order to get the double essence from them," because each of these two approaches "leads to the other; they form a circle, and there can be no other center to the circle but the empirical study of evolution" (115). The circle thus drawn demands a paradigm shift, a change in perspective. It is not a question of putting both methods on the same plane, but of constructing an assemblage dominated by the time of creation. It is in this sense that we can speak of a materialism of the event.

Similarly, it is not a question of opposing fabrication to creation, since it is through fabrication that creation is rendered possible, even though there is a difference in kind between the two.[45] In the fabrication of apparatuses for the completion of teleological action, the intellect goes beyond the intended object, which gives rise to a disinterested type of work. That is, it releases something other than the mere completion of the purpose that had been intended. "Though we derive an immediate advantage from the thing made, as an intelligent animal might do, and though this advantage be all the inventor sought; it is a slight matter compared with the new ideas and new feelings that the invention may give rise to in every direction, as if the essential part of the effect were to raise us above ourselves and enlarge our horizon" (Bergson, *Creative Evolution*, 118).

Fabrication, which operates according to a plan and a purpose, creates the conditions for consciousness to distract itself and avoid being completely absorbed in the completion of teleological action, thus exempting it from focusing exclusively on material bodies whose flow would initially lead it and then engulf it. The intellect starts by fabricating instruments. This fabrication is only possible through the use of certain means that are not cut to the precise dimensions of their object, that exceed it, and thus allow the intellect to engage in a supplementary, so-called disinterested, type of work. Like the relationship between consciousness and intellect, it is impossible to say that fabrication is the *cause* of creation but only that, through it, the process begins. Given these conditions, we can finally understand fabrication as invention—the origin of industry—whose true nature the intellect fails to perceive. And we can finally understand invention as a creative act that participates in the production of being "in its *upspringing*, that is to say, in its indivisibility,. . . in its *fervor*, that is to say, in its creativeness" (106).

Finally, we can understand fabrication not as work but as the force of invention. However, in capitalism the liberation of the possibility of choice has reached such a degree that it cannot be compared to the ahistorical fabrication described by Bergson. This is the very definition of capitalism: creation is no longer related to a purpose, to a use value, but to itself. In capitalism, to speak like Bergson, creation grasps and reveals itself materially, as a growth from within the uninterrupted continuation of the past into a present that encroaches upon the future.

When Bergson wants to describe, through an example, how fabrication is surpassed by creation, he is compelled to refer to capitalist industry and the machines that organize it. Thus he describes the first steam machine, designed by Thomas Newcomen, which required the presence of a child whose sole responsibility was maneuvering the taps. From the moment a child invented an automatic way of opening and closing taps, we witnessed phenomena that had a radically different scope. If we look at machines, we only see a "slight difference of complexity." But if we look at the children employed, we "see that while one is wholly taken up by the watching, the other is free to go and play as he chooses, and that, from this point of view, the difference between the two machines is radical, the first holding the attention captive, the second setting it at liberty" (119).

This force of invention is radically distinct from fabrication, because it had to pass through matter and the satisfaction of needs (necessity) in order to close the oppositional gaps between the intellect and consciousness, the body and mind, the intelligible and sensible. By squeezing these oppositions more closely together, Bergson is able to discover their common source. The deterritorialization of matter into flows rather than things,[46] but also electronic and digital technologies as well as the continuum they trace, is a motor for the production of duration and affective force that "prepare[s] the way for a reconciliation between the unextended and the extended."[47] "We are now, then, able to attempt a genesis of intellect at the same time as a genesis of material bodies" (Bergson, *Creative Evolution*, 120).

Therefore, I can say that the task is not—as Bergson believed, according to a strange theory of "mechanical mysticism"—the definition of a new mind (*esprit*) for a body that has been disproportionally developed within capitalism, but the conception of a process for the reciprocal constitution of intellectuality and materiality on the basis of this common source: the force of time. It is only on this condition that we can understand how these technologies intervene in a completely different way than mechanical and thermodynamic technologies in the cocreation of the real, in the world in the making.

2.5.4

The relationship between the order of the vital, or willed, and the order of the inert, or automatic, is clearly developed in the case of the constitution of the body. According to Bergson, the increasing complexity of the organism results from the need to complicate the nervous system. As we know, the greater the complexity of the brain and nervous system, the greater the interval between action and reaction. And what does this complication consist in? In a simultaneous development of automatic and voluntary activity. There is no opposition between the two orders of development, since the automatic provides an "appropriate instrument" to the will. An organism is a series of motor mechanisms that the will uses to construct itself or chooses to initiate. "The will of an animal is the more effective and the more intense, the greater the number of the mechanisms it can choose from, the more complicated the switchboard on which all the motor paths cross, or, in other words, the more developed its brain" (162).

It is through the montage of mechanisms that "the progress of the nervous system assures to the act increasing precision, increasing variety, increasing efficiency and independence" (162). The body and its mechanisms are therefore functions of the capacity to act. It is action and will that construct the body and the organism, and not the reverse. The development of electronic and digital technologies is effectively an automation of perception, memory, and imagination. But this automation is developed to make action more precise and effective. Consequentially, it is not surprising that the organism is also subject to metamorphoses that go beyond the human condition, since each will to power constructs its own body. "The organism behaves more and more like a machine for action, which reconstructs itself entirely for every new act, as if it were made of india-rubber and could, at any moment, change the shape of all its parts" (162).

Bergson distinguishes between living bodies, isolated and enclosed by nature, and inert bodies, carved out by our perception according to the interest of action, which is itself guided by virtual bodies that aim to constitute themselves. Living bodies and inert bodies thus "determine one another by a semi-artificial operation entirely relative to our future action on things" (146).[48] Contrary to what cyberculture believes, this semiartificial operation is not specific to digital technologies but is proper to any creative process. Technologies of time merely increase the force of this semiartificial operation.

3

VIDEO, FLOWS, AND REAL TIME

3.1 RECORDING (OR HABIT)

The mechanical eye, the camera, rejecting the human eye as crib sheet, gropes its way through the chaos of visual events, letting itself be drawn or repelled by movement, probing, as it goes, the path of its own movement. It experiments, distending time, dissecting movement, or, in a contrary fashion, absorbing time within itself.

—Dziga Vertov

3.1.1

If cinema has revealed that the world is a flow of images and that the world of images is in continuous variation, video technology initiates a further deterritorialization of these flows. It reveals not only the movements, the infinite variation of images, but also the time-matter of which these images are made: electromagnetic waves. Video technology is a machinic assemblage that establishes a relationship between asignifying flows (waves) and signifying flows (images). It is the first technical means of image production that corresponds to the generalized decoding of flows.[1] Photography is already a technology that crystallizes time, since the image it creates is connected to the camera's shutter speed and its capacity to retain time. Photography seizes a becoming by fixing it. Cinema, by unrolling the photogram, gives the illusion of movement, according to Henri Bergson's definition.[2] But only video technology manages to capture movement—not simply the movement through space, but the pure vibrations of light.

> In video, movement is light. It is movements, before anything else, that make up the structure of the video image. It is much more in the structure of the image than in the movement itself, that is, in the image that traverses space. Video is directly associated with light since it is a technological transformation-codification. Movement is produced by the electronic structure of the image: its grains, its lines, its frame. In objects there are movements, there are frequencies, there are atoms, there is energy. Video makes possible the perception of these energetic objects and thus the discovery of a different reality.[3]

The genetic element of cinema is still the photogram—at the technological level of image creation, since montage will introduce another genetic element, a temporal element—while in video it is time.[4] Cinema technology, from this point of view, corresponds to a moment of transition toward a generalized decoding of flows: its technology is an assemblage of photographic impressions—chemical imprints of light upon a support—and flows, the scrolling of the photogram. The automatic production of the image is not yet the product of any electronic flow, it does not yet extract from the infinite variation of asignifying figures, and it does not yet plunge into image-matter. Video technology is a good metaphor for the relationship between matter, as Bergson understands it, and the perception of the body. The video image is not a stationary photogram set in motion by a mechanical assemblage, but an image in continuous formation painted by an electronic brush. It gets its movement directly from the undulation of matter. In fact, it is this very undulation. Video technology is a modulation of flows, and its image is nothing but a relation of flows. It is contraction-dilation of time-matter.[5] And this image of raw perception comes before memory (montage), not as an icon, but as a frame of points and lines. Thus video does not present images but simply shows the weaving of lines. The difference between weaving and video is that video keeps weaving and reweaving to infinity, according to new motifs.

This process of synthesis is even present, albeit in a disguised way, in Marshall McLuhan's work. The video image projects about three million pixels per second to the viewer, who can only retain a few dozen at a time, with which he will construct an image. Every video image is a mosaic of clear and less clear pixels, and it forces us in each moment to synthesize the elements of this mosaic in an intense participation of all the senses.

3.1.2

In this regard, the Bergsonian characteristics of video, as well as its specific relation to cinema, had already been defined by video artists in the 1960s. The cinema camera is still too close to the illusion of perception as the impression of an object on a support, whereas it is sufficient to simply connect a video camera in order to see images. "You don't need a recorder to have video. You turn it on, and the circuits are all activated—it's humming, it's going . . . it's all connected—a living, dynamic system, an energy field."⁶ We are plunged into pure vibrations, into the circulation of time-matter. The decision to record consists in turning on the recorder, not the camera. "The camera is always on, there is always an image. This duration, this always-there, can be said to be real time."⁷ When one makes a video, one plugs into and interferes with the continuous process of universal variation that exists prior to any intention of working with it; one installs oneself in the flow. This duration can be called "real time," a duration that cinema does not know. Television has rendered visible the characteristic of video in which the proliferation of images is infinite. With television, the world has always been made up of images. It is therefore no longer necessary to represent or create images. As the artists tell us, we are "working with images" not "creating images."⁸

3.1.3

In order to understand the specific differences between the technological assemblages of cinema and video, further analysis is necessary. A rapprochement between cinema and Bergson's interpretation of it has already been brilliantly conducted by Gilles Deleuze. In fact, it is thanks to his work that my reflection on this topic was able to assume consistency. Deleuze makes a very original and relevant comparison between the concept of pure perception and the fact that cinema returns the world to us as a world of images. Starting from these insights, I would like to illustrate not only how video technology returns the world to us as image but also that it reproduces the relationship between perception and memory, as theorized by Bergson. In this regard, I compare the way that Nam June Paik discusses the production of color by video technology with Bergson's analysis of the apparatus of color perception in consciousness.

We have defined pure perception as an image, but it must be remembered it is the image in itself, the image that is not seen by any eye. It would be more accurate to speak about the flow of light or even, as Bergson suggests, to make this concept perceptible, through an image, to our consciousness of pure, infinite vibrations. The image itself is already a synthesis, a fixation of pure perception. It is a selection and contraction of pure vibrations, performed according to the needs of action. "The truth is that this independent image is a late and artificial product of the mind."[9] To use a beautiful Bergsonian metaphor, the image in perception, like a video image, emerges from visual dust.

Video allows us to go beyond the image, to access something of the dimension of pure perception: flows of light, matter-flows. It can venture to places that our consciousness cannot, accessing and working on something like Bergsonian pure perception. According to Deleuze, pure perception exists for us only in principle. It is therefore an abstraction, since it is inseparable from a filter through which things emerge. "A great screen has to be placed in between them. Like a formless elastic membrane, an electromagnetic field . . . it makes something issue from chaos. . . . From a physical point of view, chaos would be a universal giddiness, the sum of all possible perceptions being infinitesimal or infinitely minute; but the screen would extract differentials that could be integrated in ordered perceptions."[10] According to Deleuze, there are an infinity of filters or superimposed screens, from our senses to the ultimate filter beyond which there would be pure perception or chaos. The electromagnetic screen of video is a filter that is closer to pure perception than the filter of our senses.

The difficulty in thinking pure perception lies in the fact that we must abandon the homogeneous categories of space and time with which we are accustomed to think and see. If we abandon these categories—which correspond to the schemas of our power to act and to the conditions of our faculty of understanding, but not to the qualities of things—we enter into another dimension, that of perception-matter. A world of extensive and intensive forces, a mixture of extended and unextended, of quality and quantity.

Bergson's work functions like one of Paik's synthesizers; that is, with the capacity to reconstruct a continuity between beings similar to the unity of human and nature, which offers a new conception of matter and a new power of metamorphosis and creation. The definition of pure perception as a dimension beyond our categories of space and time—constituted by a mixture of extended and unextended, quality and quantity—seems, at first glance, difficult to understand. But new technologies and science allow us to see something of this world.

3.1.4

According to Bergson, our concrete perception is the instantaneous contraction of infinite vibrations; a synthesis of the flow of pure vibrations of perception-matter. What is a sensation or an image? It is the operation of contracting trillions of vibrations on a receptive surface, trillions of little shocks. The image is therefore a contraction of flows. "The qualitative heterogeneity of our successive perceptions of the universe results from the fact that each, in itself, extends over a certain depth of duration and that memory condenses in each an enormous multiplicity of vibrations which appear to us all at once, although they are successive."[11] Perception, as either image or sensation, is located, according to Bergson, at the confluence of consciousness and matter. It condenses, within a duration that is proper to us and that characterizes our consciousness, immense periods that

Bergson defines as the "duration of things" themselves. We instantaneously condense an extremely long history that occurs in the outside world.

Let us take the example of red light, which has the longest wavelength and whose vibrations are therefore the least frequent. In an instant, it undergoes four hundred trillion successive vibrations. What we perceive as red light is a division and contraction of duration according to the capacity of contraction-relaxation in our own duration, which is not defined by physiology but by the power of action. Bergson often says that perception ceases at the point where our capacity to act does. Under these conditions, perception is a relation of time, of duration. By utilizing one's category of understanding, time, the duration of pure perception, is confused with space and the duration of human action. The role of the body and mind is, in different ways, to connect the successive moments of the duration of things in a time and space in which one can act; that is, to contract the duration of things into a human duration. "May we not conceive, for instance, that the irreducibility of two perceived colors is due mainly to the narrow duration into which are contracted the billions of vibrations which they execute in one of our moments? If we could stretch out this duration, that is to say, live it at a slower rhythm, should we not, as the rhythm slowed down, see these colors pale and lengthen into successive impressions, still colored, no doubt, but nearer and nearer to coincidence with pure vibrations?"[12]

Bergson calculates that in order to see the duration "in itself" of an instant of the color red, given the rhythm of our consciousness, it would take 250 centuries. Memory therefore performs a contraction of pure duration, which is capable of storing an infinite number of phenomena, into a human duration that divides and solidifies it. This division is done spatially. Indeed, Bergson asserts that our inability to conceive of perception as a relationship between temporalities is due to our habit of relating all movements to space. Our perception always inhabits a homogeneous space within the indefinite multiplicity of matter. But we must instead consider the movements in time. As we shall see, this is exactly what video technology does. From this perspective, video is more veracious than the natural perception whose disappearance we mourn.

3.1.5

Now we can fully understand Nam June Paik's mysterious affirmation that video is time. According to him, the technological assemblage of video imitates the relationship between the different temporalities of which Bergson speaks. How does video produce color? By modulating, in a specific manner, the flow of matter through a technology that deals with the becoming of this flow. The video machine functions exactly like the human brain, translating a movement that is imperceptible to our categories of space and time into another movement that is perceptible. The pure perception of video, its matter-energy, is constituted by electromagnetic waves that are the pure vibrations from which images are constructed. Color is an electromagnetic wave made up of specific vibrations that are contracted by the video machine into a duration suitable to humans. But in this case, it is technology that functions like a mind or subject in order to reduce the duration of pure perception into a human duration. It is the technological assemblage that organizes the relationship between flows. Just listen to Paik:

> Since televisual space does not exist, all spatial information has to be translated into lines and points without thickness.

Therefore the signal can be transmitted wirelessly, on a single channel. They also had to put all the colors on that line. To do this, they invented a sort of social contract. A wave, known as the chromatic carrier, is 3.5 millionths of a second long. Although they are already very small, these waves are again divided into many phases, for example, seven phases representing the colors of the rainbow. The first seventh of this wave is called "blue," the next seventh is called "yellow," the next "orange," and so on. This circuit opens and closes very quickly—21 million times per second—passing through the colors in order. As in nature, it is the very, very rapid temporal succession that produces color in television. It's a social contract. When you make a movie, nature colors the film strip through the lens. But in television, there is no direct relationship between reality and images, only code-systems. *We therefore enter the temporal dimension.*[13]

The video image is a contraction-modulation of flows of light. This relation is not determined in the machine by the memory, but by a technological apparatus. Video perception is closer to the apparatus described by Bergson than to the physiology of the eye. In fact, it is a technology that contracts perception-matter, movement-matter, wave-matter, and infinite vibrations within a human duration. Color is restored to our perception through a technological treatment of duration and time. Natural perception is ultimately a particular transformation of the asignifying fluxes that, through the intervention of the memory and the brain, are rendered perceptible. Video technology imitates this relationship between flows and consciousness. We also find hints of Bergsonian metaphysics in the relationship that Bill Viola defines between human perception and the perception of the video camera. Here the transition from pure perception, with its vibrations and frequencies, to the image defined as a "division in time" highlights the specific differences and limits of human perception.

> The video image is a standing wave pattern of electrical energy, a vibrating system composed of specific frequencies as one would expect to find in any resonating object. As has been described many times, the image we see on the surface of the cathode ray tube is the trace of a single moving focused point of light from a stream of electrons striking the screen from behind, causing its phosphor coated surface to glow. In video, a still image does not exist, in fact at any given moment a complete image does not exist at all. The fabric of all video images, moving or still, is the activated constantly sweeping electron beam—the steady stream of electrical impulses coming from the camera or video recorder driving it. The divisions into lines and frames are solely divisions in time, the opening and closing of temporal windows that demarcate periods of activity within the flowing stream of electrons. Thus, the video image is a living dynamic energy field, a vibration appearing solid only because it exceeds our ability to discern such fine slices of time.[14]

3.2 MONTAGE: PROCESSING (OR MEMORY)

The school of kino-eye calls for construction of the film-object upon "intervals," that is, upon the movement between images.

Intervals (the transitions from one movement to another), and not the movements themselves, are the material, the elements of the art of movement.

Editing tables containing definite calculations, similar to systems of musical notation, as well as studies in rhythm, "intervals," etc., exist.

—Dziga Vertov

3.2.1

As I have just described, shooting with video can be related, by analogy, to the sensory-motor function of the human body described by Bergson. Indeed, the technological apparatus of video only transforms one movement into another, even if the possibilities of contraction-relaxation are much more numerous than those of the body. They operate solely on the plane of the present. However, image processing techniques make possible the reproduction of the "free" labor of memory.[15] Just as raw perception is reworked by the activity of synthesis in intellectual labor, video recording and editing allows for the infinite production of images. Technology increases the power of intellectual labor, but it is only with video that the first step is taken to endow the machine with a memory that resonates with the actual object and thus reproduces the circuit between the actual image and memory. Image processing technologies are syntheses of flows that introduce a degree of freedom into the treatment of durations through shooting. On the one hand, video renders the world of pure perception, with all its virtualities and actualities, accessible to humans, and on the other hand, the contraction and relaxation of time-matter finds in editing technology almost infinite possibilities of creation. That is, the relationship between durations—human duration and other durations in the universe—reveals assemblages that allow us to go beyond human forms of experience and representation.

For Nam June Paik, only image processing allows for the introduction of true memory into video. For him, the video camera is simply an "input-time" and "output-time" apparatus that is inserted into flows (waves) of light, an apparatus without freedom that reproduces contractions and relaxations in the form of habit. If we limit ourselves to the apparatus of the camera, we remain within the present; that is, within the simple contraction-relaxation of time-matter and perception-matter. But in order to contract and relax the time of memory, technologies of image processing should be implemented.

> However in our real life—say, live life—the relationship of input-time and output-time is much more complex—e.g., in some extreme situations or in dreams our whole life can be experienced as a flashback compressed into a split second (the survivors from air crashes or ski accidents tell of it often) . . . or, as in the example of Proust, one can brood over a brief childhood experience practically all of one's life in the isolation of a cork-lined room. That means, certain input-time can be extended or compressed in output-time at will . . . and this metamorphosis (not only in quantity, but also in quality) is the very function of our brain which is, in computer terms, the central processing unit itself. The painstaking process of editing is nothing but the simulation of this brain function.[16]

More precisely, we could say, with Bergson, that video montage simulates memory and intellectual labor rather than "material syntheses." The movements of extension and compression that Paik mentions operate with the crystallized duration of the video camera. Once again, the difference between video and film montage is remarkable.[17]

Viola defines the transition from video to simulation and digital technologies as the development of the relationship between perception and memory, since the video recorder is already a simulation of intellectual labor, even if only in a crude form. "After the first video camera, with its recorder, gave us an eye connected to a coarse form of nonselective memory, we are now in the next stage of evolution: the era of artificial perception and intelligence."[18] By claiming that simulation technologies progressively reduce the crudeness of the work of memory as they develop, Viola allows us to retrace the machinic phyla of different technologies in a new way.

3.2.2

If cinema reaches the end of its development in the time-image—solely through aesthetic procedures—in which the mind makes of all motor movements of bodies a direct experience of past time, as far as video is concerned, this experience is inscribed in its very technological operation. The essential concept for video is time and not movement. It is time that is intrinsic to video and not vision, as implied by the etymological root of *video*. The essential capital of video resides, as Viola suggests, not in "chronological time but in a movement that is contained within thought, a topology of time that has become accessible."[19] As we have just seen, this happens in three different ways: the contraction of time-matter, the synthesis of the past, and the conservation and accumulation of time in order to intervene into time in the making. This capacity to intervene into time, *to retain time in order to intervene into the durations of the world* and act upon the present in the making—is the "live" quality of video.

"While Plato said that art imitates nature, video imitates time," according to one of Paik's formulations. Video is the first technology that imitates the various functions and syntheses of time. Video technology is not a temporal technology simply because it modulates time-matter, but also because it always works on a duration. That is, unlike the technology of cinema, it exists, strictly speaking, only immediately, in the event. In cinema, time is by definition a delayed time—in which we can represent time and its syntheses—while electronic and digital systems exist in the real time of the production of social time, its overflowing and continuous renewal. The only people who have taught us anything about this technology—artists—never compare it, or otherwise only negatively, with cinema, precisely because they work in real time. "From an existential-technological point of view, we are close to the telephone and the radar screen, which require that we respond, otherwise communication is not only interrupted, but it has not even begun."[20] It cannot be emphasized enough that the real time of video technology is completely different from the real time of television. The technological machine is often confused with the apparatus of power. Television is an apparatus of power, which is constituted precisely as the denial and diversion of the ontological consistency inherent to video: real time, time in the making, time that passes and splits. This time, it must be stressed, is the indeterminate time of creation, choice, and event.

Soviet filmmakers, beyond all the mystifications they are subjected to, immediately grasped the specificity of television in its capacity to perceive the event in the making. Video makes it possible to simultaneously use what cinema must separate: shooting and montage, perception and memory. The result of combining shooting and montage within the same process is that it becomes possible to capture the movement of time (and thought) in the very moment of its outpouring. Already in 1958, in his *Notes of a Film Director*, Sergei Eisenstein described television as a "new hope" for thought:

> Then there is the miracle of television—a living reality staring us in the face, ready to nullify the experience of the silent and sound cinema, which itself has not yet been fully assimilated. There, montage, for instance, was a mere sequence (more or less perfect) of the real course of events as seen and creatively reflected through the consciousness and emotions of an artist. Here, it will be the course of events itself, presented the moment they occur. This will be an astonishing meeting of two extremes. The first link in the chain of the developing forms of histrionics is the actor, the mime. Conveying to his audiences the ideas and emotions he experiences at that moment, he will hold his hand out to the exponent of the highest form of future histrionics—the TV magician—who quick as a flash will expertly use camera eyes and angles to enthrall the millions-strong TV audiences with his artistic interpretation of an event taking place at that very moment.[21]

Eisenstein shows us that with video we are no longer in the regime of representation. Re-presentation presupposes an encounter between two temporalities, which might be very far from each other: the present of the artist and the present of the one receiving the work (writing, painting, film, etc.). The temporality of video is, however, a present that returns us to the time of the event, to time in the making, which simultaneously implies the artist and the spectator, an open duration that can provoke a reversibility between creation and reception. Video therefore calls for other aesthetic powers and creative assemblages, but also other forms of communication.

3.3 REAL TIME (OR THE SPLITTING OF TIME)

3.3.1

From the human eye's viewpoint I haven't really the right to "edit in" myself beside those who are seated in this hall, for instance. Yet in kino-eye space, I can edit myself not only sitting here beside you, but in various parts of the globe. It would be absurd to create obstacles such as walls and distance for kino-eye. In anticipation of television it should be clear that such "vision-at-a-distance" is possible in film-montage. The idea that truth is only what is seen by the human eye . . . is refuted by the very nature of man's thought.

—Dziga Vertov

The concept of real time may lead to misunderstandings about the temporalities of video and digital technologies, since it refers only to the simultaneity and immediacy of the flow of information, which we see, for example, in the work of Paul Virilio. However, one of Walter Benjamin's comments on Bergson's *Matter and Memory* reminds us that acceleration is not an exclusive characteristic of technology, and that the first instruments of acceleration were memory and perception.[22] "Within the space of a second . . . to perceive means to immobilize."[23] Thus Bergson presents human perception as the first instrument of acceleration. Moreover, the capacity to accelerate time is, in his view, a fundamental condition of human perception. This analysis is radically opposed to Virilio's position. The acceleration of time increases our capacity to perceive rather than diminish it, because acceleration is an operation that allows us to contract more reality within the same instant. This synthesis intensively increases our possibilities for acting in time. It increases our capacity for delaying, for indetermination, and therefore, for choice.

According to Jean-Louis Weissberg, Virilio confuses the time of transportation and communication with the time of image processing. The former accelerates, but the second increases exponentially. What increases is the possibility of delaying and therefore of introducing indetermination and the unpredictable into the real. The capacity to work time, to retain it, is in no way diminished. The problem lies elsewhere: the power of image processing—creative power—has been withdrawn from social practice. It is strictly controlled and codified by both the state and the media. We have only our individual and isolated subjectivities to process these speeds and images.

This is evident in the case of television, but also with regard to new digital technologies and their integration into online networks. Multinationals in the communication industry try to control and impose the technological standards of image production and communication. Virilio completely sidesteps this political problem and therefore introduces a mystification about the nature of these technologies, because while the speed of image transportation (information) is indeed accelerated, the processing of these images (and information) renders them infinitely variable. He naturalizes perception and spatializes speed and time. His critique of the industrialization of vision and perception thus loses its force, since it always confuses the ontological plane of these technologies with the apparatuses of power that assemble them according to their own ends.

In reality, when artists claim that video is time, they are referring to a nonchronological temporality.[24] Technologies of vision liberate us from natural perception, its illusions and its anthropocentrism, and thus make us enter into other temporalities. They liberate us from the subordination of time to movement and open us up to a direct experience of time. This movement liberated from physical motion is intensive; it is nonchronological time, the time of the event. Capable of prolonging the past into the present and of opening onto the future, this movement is a present that is not a mathematical point but rather the coexistence of past and present and their continuous unfolding. It is the time that causes movement to arise, that opens up new bifurcations and virtualities. Here, the instant is a becoming, which instead of being passively encased between the past and the future, becomes germinative and develops ontological coordinates.

Cinema, the first experiment in which movement is liberated from physical motion, reaches the "representation" of the emergence of time, a creative and ontological time. The time of video is significantly different from that of cinematic time, because here we are in time, we participate in the construction of the event. We do not "see" the time of video; we live it. The political importance of this temporality requires development at greater length.

3.3.2

The limitations of most of the critiques of television consist in the fact that they are stated from the perspective of cinema, from the regime of the image that is specific to cinema. This perspective risks missing the specific ontological consistency, the differences, and the centers of subjectivation specific to video. In short, with video we have entered into another regime of the image and another temporality that requires its own conceptualization. It is true that the ontological consistency of video is difficult to locate: from the beginning, video has been a technology overcoded by the state. It is therefore necessary to listen, once again, to video artists in order to identify the aesthetic and social potentialities of video. The most perceptive critics of television denounce the "end of the adventure of perception" (Serge Daney) or the "social and cultural function" (Jean-Luc Godard) of television mostly with a nostalgic regret for the lost aesthetic function of the cinematic image. But if there is nothing to see "behind" the image, if there is nothing more to see "in" the image, it may be because the video image provokes other forces and practices. If the video image no longer makes us perceive the representation of time as such, it may be because it has itself become that time, a place, a space of action for time as such. It is no longer just an image to be seen, but an image on which one intervenes, on which one works (a time of the event).

As Paik claims, the camera is nothing more than an "input-output unit" caught up in and traversed by flows. It works on the temporality of these flows, in real time. The specificity of the video camera is to receive inputs and to return outputs in real time. All the work of video occurs between this input and output: connect to a flow, work it, transform it, and return it to circulation to be worked again. The video image is an image to be touched rather than seen. With it, we intervene in incorporeal matter more like a painter or sculptor than a filmmaker. We can no longer remain seers—in the best case, as the cinephile would like—or voyeurs—in the worst case, as television would like—in front of the video image. And the video image largely anticipates the virtual image. We often mock, and rightly so, the social and cultural function of television. But is there not, perhaps, a function that cinema can never fulfill and that has hitherto only been exploited by apparatuses of power? For example, consider liveness, a specific ontological consistency of video. "Live TV is like life, you never manage to get everything done, there are always a few bittersweet regrets.... But what is the interest in being live? Interactive events must happen live. We'd like to use interactive TV because the best use you can get from a TV is to answer it like a telephone."[25]

It is not simply a matter of affirming the communicative, social, and cultural functions of video but, on the contrary, of defining another temporality. The technological assemblage of video does not show us the time of the event but puts us in the event. With video it is not about supplementing time but constructing it and doing so collectively, in an assemblage, in a flow. Live technologies impose a concept of subjectivity that constitutes a virtual critique of the concept of the spectator, formulated by Joseph Beuys in this way: "I am a transmitter and I radiate." By this he means not only that he is in the flow but that he himself is a flow (fluxus), connected to other flows, infinite variations of them. This implies a completely different social and aesthetic practice that is inextricably linked to flows, time, and multiplicity.

Time is not given in the image, but constructed in a situation. This is the great innovation of video: a machinic assemblage of situations and not of vision. Perhaps it is in this sense that we should interpret another statement from Paik: "The ultimate goal of the video revolution is the establishment of space-to-space communication, from situation to situation and not from subject to subject ... we no longer work on an object, but a situation."[26] We work on any situation whatsoever. It is for this reason that, while projection rooms have broken out around the world, it is only with video that the world properly becomes a world of images in the Bergsonian sense. The world creates its "cinema," but only with video technology.

Can we find new possibilities of action in video, in which we no longer have the image prolonged in action but the image itself as action? The first artists who used video to create performances, and not for projection, certainly understood these truths. Power, after separating assemblages and flows from their virtuality, from their relationship with the outside, after imprisoning flows within a vicious circle—the power that demonstrates power—has made its event, the always equal and infinite reproduction of its becoming, a becoming of the same, a becoming that is not a metamorphosis, but a reproduction. "The best part of [John] Cage's work is his live electronic music, an art form linked to both time and space that can never be transferred onto a video or audio disc. The 'great art' of video will come in the form of video installation and a genre of notation will be developed that will allow the transmission of certain types of works of art."[27]

The space-time blocks of video are directly embedded in life; they are the virtuality that accompanies "live" life. The nostalgia for cinema is a nostalgia for the past. We need a nostalgia for the future. Why didn't anyone follow Roberto Rossellini on this path? It was obviously not a question of the social or the cultural—as opposed to art—but of grasping the ontological and aesthetic consistencies of this new technology, of working and going along with them. These aesthetic functions can only be developed if they are immersed into a flow, an assemblage, an action, an event. This immersion into multiplicity is not delayed or recorded, but live. Television, as an apparatus of power, can construct its events only because it understood, better and faster than its critics, the ontological consistency of the video image. This new regime of the image only began to be understood with the appearance of the virtual image; that is, when it entered into other assemblages.

3.3.3

Serge Daney has claimed that television does not exist. This statement can only be grasped in the sense in which Michel Foucault claimed that power does not exist. That is, it does not exist in the form in which it has been understood until now, and to critique it we must reconstruct its function according to other categories and assemblages. Therefore, television does not exist as a propaganda machine or an aesthetic assemblage, but it does exist as an ontological (temporal) condition of our society. Television is the guardian of time. Television functions as social memory in the two forms we know of the constitution and conservation of time. Bergson does not take this type of memory into consideration—though Benjamin partially analyzes it—but it can be defined using his concepts.

The spectacular development of television is contemporary with the domination of the temporality of capital over society as a whole. Noncapitalist temporalities that have, for a long time, survived alongside the time of capital have been destroyed and have lost any capacity to determine human experience. Benjamin discusses this. But as Antonio Negri has demonstrated, it is around the problem of time—the opposition between power-time and value-time—that conflict is redefined in post-Fordist capitalism.[28] Indeed, this conflict presents an empty time in the face of a power-time, a duration that, as Bergson puts it, means invention, creation, and the continuous development of the absolutely new. Capitalist deterritorialization has liberated the hidden ground of time—the virtual/actual circuit—from the repetition of the present (habit/custom) and the repetition of the past (memory/tradition). It is this heterogeneity of time, this duplication, this source of continuous creation, that television must control, neutralize, contain, and solicit in the direction of chronological real time, which is only the repetition of what exists. "Time and its indeterminacy" could be written in all television credits, because television is about this and not manipulation.

Since memory is the extension of the past into the present, the capacity to interpose the past into the present with a view toward action, television must constitute our past in order to interpose it in the production of subjectivity. As we know from Bergson, the construction of memory, of the past, is done in the present, since the past must be contemporary with the present. It is for this reason that television is obsessed with the news (*actualité*), such that it has to constantly film it, duplicate it, with its images. Media coverage—and it is indeed a matter of covering immediate perception and actuality with a layer of memory-images—is obviously a selection, preserving what is useful for the reproduction of power and its representation, because it decides what must become collective memory. The splitting of time, its contemporary and incessant production of virtuality, must be controlled and neutralized by television images. It is this imperative to cover the news and actuality with a shroud of memory-images that explains the need to film everything.

The social images of our collective memory are images from television. When television decides that there will be no pictures (for example, during the Gulf War), this is because it does not want to have indefinite, open time. This decision causes a shock and a hole in memory, which we then attempt to fill in as best we can. It would be possible to fill it by producing other images, but at the moment this is almost impossible at the social level.[29]

The work of televisual memory has nothing to do with derealization, mystification, or propaganda. Through television, power tries to control and neutralize the social virtuality (if we can call it that) of our actuality. With this operation secured, all the others can follow: propaganda, ideology, mystification. But reversing the order of priorities would be a big mistake, prohibiting access to the ontological foundation of television.[30] This is a fundamental error of thinkers on the Left.

That information is driven by other information, that an image is driven by another image, is not scandalous but rather corresponds to the normal operation of memory. What is important is the contraction-relaxation of time as worked by memory. This is another important function of television. The contraction of memory consists in searching the memory—in memory-images, in the layers of the past—for images to cover over actuality such that they can be associated with it and therefore constitute new perceptions. Just try to remember some remarkable moment in collective history, and you will see that it is almost always an image from television. The past is imprisoned by television. In the same way, television controls the contraction-relaxation of time. It also controls the intensity with which one is situated in time. It accelerates time, slows it down, modulates it. From this point of view, the signifying level is willingly abandoned by those who still think that the media consists mostly of ideological powers.

My goal here is not, like the Frankfurt school, to denounce the totalitarianism of television. I think it functions as a "machine of capture"—capturing affective forces that overflow it—that solicits the production of the new only to control and sterilize its limits and intensity. Whether this process succeeds does not simply depend upon apparatuses of communication. The production of subjectivity specific to the media should not, therefore, be understood as an ideological process, but primarily as a capacity to control, at the social level, the relationship between perception and memory. The regulation of time, in and by which this relationship is constituted, is today insured by television. It is a duration, a rhythm that enters into relation with other durations and other rhythms. The actual-virtual circuit is the secret of how television works. Television is time, from all points of view.

3.4 THE SPECTACLE

We therefore take as the point of departure the use of the camera as a kino-eye, more perfect than the human eye, for the exploration of the chaos of visual phenomena that fills space.... The position of our bodies while observing or our perception of a certain number of features of a visual phenomenon in a given instant are by no means obligatory limitations for the camera.... We cannot improve the making of our eyes, but we can endlessly perfect the camera.

—Dziga Vertov

Why are we able, according to Gilles Deleuze, to define cinematic techniques as Bergsonian techniques? Because they make obvious the fact that there are images, in the same way that Foucault tells us that there is language. Contrary to what all the theories of representation claim, there is no longer the image and the thing, the image in consciousness and the thing in the world. Every thing is an image, and every image is a thing. The brain is only one image among others, and the subject is only a particular type of image. Therefore, perception is in things. It is not the privilege of a subject, since the subject is only a particular form of the perception-image. This implies that there is no longer any distinction between the world and images, and that the object ceases to be independent of the image that describes it. We have entered the world of the spectacle. This idea is realized by television, since it creates on the social level, for the first time, a nondistinction between the actual and the virtual by splitting, in real time, the present of a world of images. Therefore, the present is simultaneously conserved and continually renewed. We have really entered the world of the spectacle in the sense that the actual and the virtual reflect one another to infinity. Cinema merely announced and represented this new condition of collective perception, as one represents a book, a painting, or a sculpture.[31] Postwar cinema, as Deleuze marvelously demonstrated, is a cinema of the time-image in which we can see the nondistinction of the actual and the virtual, the pure and empty form of time. But it is only with television and digital technologies that, rather than represent this new dimension of the time-image, we make it live.

Cinema no longer represents the conditions of perception and collective memory. The relationships between subject, object, and spectator—as well as the space of classical representation that they determine—are completely redefined by the spectacle. This process of the deterritorialization of the stage that began with cinema has been radicalized by video. Indeed the space of representation in video escapes painting, the stage, and even the screen, as one of the first video artists of the early 1960s said in a surprising and prescient way: "The space of representation has become closer and closer to the space of the brain, and the next revolution will be to connect video directly to the brain, without a camera. Video is just a transitional technology on the way to the complete definition of the space of representation without a camera,"[32] on the way to virtual images. But here it should be made clear, with Bergson, that the production of mental images does not take place in the brain—the brain is only the seat of sensory-motor movements—and that consequently it is necessary to work on the connection with the actual-virtual circuit, which depends upon intellectual effort and memory; that is, time. The deterritorialization of the flow of images (there are, strictly speaking, only electromagnetic waves); of the screen (the screen in video is only a convention, an imposed standard); and of the space of representation (the projection room is only any space whatsoever): it is in this sense that video is close to Bergson.

4

BERGSON AND SYNTHETIC IMAGES

4.1 THE VIRTUALITIES OF VIDEO ACTUALIZED IN SYNTHETIC IMAGES

4.1.1

How can we describe the passage from electronic to digital technologies with regard to the production of images? Unlike most researchers working on these technologies, I do not think we can comprehend this passage only on the basis of the difference between analog and digital. What is important, it seems to me, is the change in the production process of retention, the accumulation of time, as I have tried to reconstruct it through the work of Bergson. According to this analysis, what marks the specificity of digital technology is the fact that in the production of images we find the imitation of syntheses that constitute mental images and mental spaces. We also apply the paradigm of the imitation of time and not nature, which we used to analyze the video image, to the analysis of the production of the virtual image. But imitating time means imitating the active and passive forces that express it. This imitation of forces, as we have seen, is already engaged—even if only in a crude way—in video technology.

Therefore, compared to most researchers, I delineate the continuity break between video and simulation technologies a bit differently. There are continuities and breaks, filiations and innovations. Indeed, many of the characteristics used to define the virtual image were already present in the video image and technology. This does not imply a linear filiation between the technologies, but rather an evental development of alterity, of the virtualities contained within each machine.

4.1.2

Even in its early days, video technology was already more than the recording of objects and colors. As Nam June Paik says, there is no direct relationship, as there was in film, between reality and image. It is through a technological code that perception-matter, the wave, is transformed or crystallized into an image. In digital technologies this code is a mathematical language. Indeed, these machines involve deterritorialization in its purest, most crystalline form, since they are only a system of alternatives: zero or one, plus or minus, this or that. It is in this sense that we can speak of the subsequent deterritorialization of flows.

From the frame of points incessantly rewoven by the electronic paintbrush of video, we move to the matrix of pixels determined by an algorithmic language. The medium of the pixel is neither matter nor light, but numbers. Each pixel is locatable, controllable, modifiable. The matrix-image, even more than the frame-image, is an image always ready to be reworked; it is a forever-closed image. From this point of view, the power to weave the image is increased exponentially compared to the video frame. According to many theorists, the synthetic image is simultaneously an actual and a virtual image—a potential image—an image that can produce an infinite number of other images. As a real phenomenon, it oscillates between the actual—the image actualized on the screen—and the virtual—a potentially infinite set of images that can be computed from the same data. This technology seems to reproduce the oscillation and resonance of the actual-virtual circuit. But in reality, these concepts do not overlap with the Bergsonian conceptualization and, as I will explain, may introduce ambiguities and misunderstandings.

Temporality, which in video was already nonchronological, appears as virtual temporality. The image conceals a time that does not flow, an open time, a time without beginning or end, a matrix time, a simulation of always-renewable and different instants that can be actualized.

Just as video is an action upon images and situations (I recall that Paik compared it to the telephone and radar rather than to cinema, therefore implying a technology that demands a response), virtual technology increases this capacity to act upon images and situations. Interactivity in effect removes the distance that separates the image from the spectator. The latter can act upon the image either by a programmed instruction from the keyboard or, even more quickly and directly, by means of an electronic pencil or a voice command. Here, the tactile characteristic of the image, of which Marshall McLuhan speaks, is fully liberated. We touch the image, but with a completely deterritorialized hand: the remote control, the mouse, or the Web browser.

Thus the image is no longer held at a distance, a screen between subject and object, but is immediately accessible and transformable. According to Edmond Couchot, apparatuses that seem to insert themselves between the matrix-image and the spectator—the keyboard, the touchscreen, and so on—no longer function as media but as almost organic extensions of the image that become one with the body of the spectator. They are the image itself. To act on them is to act on the image. So we have definitively left the position of the seer. The Bergsonian ontology that refuses to split the world and its representation, reality and image, seems to find its technological realization. There is no longer the object in space and the image in consciousness, but a continuum of images.

In technologies of simulation, affection must, at a given moment, emerge from the image through apparatuses that produce the simulation of the senses. Thus the mode of operation of digital technologies traces, between perceptions and affections, a continuum that returns us to the Bergsonian common source of the body and mind. As in Bergson, in these technologies there is differentiation, a bifurcation between matter and mind, but there is also a movement, a communication, a reciprocal constitution. The problem of materialism—how to move from an image to a sensation, from the sensible to the intelligible—seems to find a new articulation. These technologies produce both the sensible and the intelligible on the basis of a common source.

The virtual image is an example of the mixtures Bergson speaks about: subject-objects beyond the turn of human experience, mixtures of the extended and unextended, entities beyond human categories of apprehension. We have entered a third zone, which should not be conceived as the dialectical resolution of these oppositions, but rather as a multiplication of degrees of the real. These are truly mutant entities, which go hand in hand with the metamorphosis that touches the human (its becoming-image, its becoming-flow, its becoming-time). These new images allow us to see and touch the hybridization between flows of matter and flows of signs.

4.2 LIGHT IN ANALOG AND DIGITAL

4.2.1

What has just been summarized is a synthesis of the characteristics through which simulation technologies and digital technologies are usually defined. But to try to understand the role that digital and simulation machines play in the constitution of the real and of subjectivity, we must return to the Bergsonian conceptualization. This is, as I have argued, a paradigm of image production that is not based upon an optical model, contrary to the models that are generally used. The Bergsonian model is a temporal model in which the production of images is based upon the syntheses of time.

On this basis, I offer the following hypothesis: the functions of contraction and relaxation of time-matter and memory (material and spiritual syntheses) may also be useful in defining the digital production capacity of images and sound. First of all, the techniques of digital recording only increase the capacity for the contraction of time-matter. From this point of view, they are defined as an interface that, in a specific way, receives and restores movements in relation to the body and to video technology. The specificity of these technologies lies in their temporality, in the duration proper to them. Thanks to the computer clock, digital technologies can contract the smallest possible intervals of time-matter and image-matter. We find a new possibility for the crystallization of image-matter that leads us further on our journey beyond the turn of human experience. Digital technologies reproduce, with even more fidelity, the work of the contraction-relaxation of memory. Second, the massive productivity of these technologies is due to the fact that their operations imitate what Bergson calls intellectual labor. This imitation operates as a crystallization of the intensive movements of the soul. The production of images and sounds, which is in no way the reproduction of existing images and sounds, is merely the result of this capacity to imitate the circuit of resonance between the actual image and the virtual image. From this point of view, it is the reproduction of intellectual labor that qualifies these new technologies. Any other qualification, however precise, is subordinate to the capacity to simulate the power contained within the actual-virtual circuit.

Most commentators reduce the rupture that technologies of simulation introduce, in relation to cinema and video, to two basic arguments. The first has to do with the fact that light plays no role in the construction of images. The second is that images are the product of a language, even if this is a programming language. But it seems that these statements, certainly correct in themselves, can be interpreted in substantially different ways, depending on the theoretical framework into which they are placed. The Bergsonian conceptualization seems, once again, to be the most useful in defining a theoretical frame of reference since it *always subordinates the optical model of the production of images to the temporal model* in which the forces of time act.

4.2.2

I will now compare Edmond Couchot's important research on virtual reality with the Bergsonian perspective. According to Couchot, the traditional image machine is characterized by the fact that it uses recording processes that are based on the treatment of light and thus work with the luminous trace. What this machine reveals exists prior to it, such that the optical image always returns to us a fully actualized and complete reality, "literally crystallized in the grain of the film or in the orientation of the magnetic particles of magnetic tapes."[1] With the information machine, according to Couchot, we see a mode of figuration that breaks radically with optical representation. The synthetic image does not reveal an optical trace, the recording of something that was and is no longer, but rather a logical-mathematical model that describes not only the phenomenal aspect of reality but also the laws that govern it. "That which preexists the image is not the object, the actualized real; it is the model—obviously an incomplete approximation of the real—its formalized description, the pure symbol."[2]

The new image no longer bears witness to the real through the inscription of light but henceforth to an interpretation of the real, elaborated and filtered by language. This position, common among researchers, is typically based on a highly questionable understanding of the "natural" processes of image production. The metaphor that is directly or indirectly used in constructing a genealogy of the development of image technologies is that of the impression of an object onto a medium. In general, this conception is developed from a definition of vision as optical perception and based upon a metaphysics that opposes world and image.

However, Bergson offers a completely different model. The role of light is certainly important in the process of constructing an image, but only as a mechanical presupposition of its production. What is most fundamental in the process of creating images is the activity of memory and intellectual labor. As we move from simple automatic recognition toward the intellectual and creative faculties, light "disappears" from the process of image production. As Bergson reminds us, imagining is not the same thing as perceiving. Indeed, light plays a completely different role depending on the model in which it operates. For Bergson, things in themselves are light; objects and sensations, as we have seen, are contractions of light. The eye, before being a mechanism for vision, is itself contracted light. If everything is light and movement, images are determined by the different syntheses of time. Only the active and passive syntheses of time enable us to see: light as such will never be able to construct images. According to Bergson, the impression of light on our brain does not, strictly speaking, produce images, but rather movements, excitations, and vibrations that affect it.[3] The image, as I understand it, requires the intervention of memory, or intellectual labor. We see and perceive only through intellectual labor. Bergson makes an interesting remark when he states that if the image we have of some object is a recording, then we have not one image but an infinity of images. This is because the image in itself is composed of millions of images, millions of vibrations.

However, most commentators limit the evolution of the machinic phylum of these technologies in two different ways: they reduce their functions to the recording of the real, and they pass over in silence the (fundamental) possibility that these technologies have to simulate the labor of memory and intellectual effort. Intellectual labor (and memory) does not appear only in the technologies of simulation but also, to varying degrees, in cinema and video. To reduce the technologies of cinema and, especially, video to recording is equivalent to limiting human perception to automatic recognition, to the perception of an automaton, or to that of an object. But this is precisely what most commentators do: they reduce the complexity of the labor of image production—which is fundamentally intellectual labor—to an optical model. Bergson, however, opens up a completely different horizon that allows us to take another look at technologies that simulate the assemblages that contribute to the constitution of subjectivity.

4.3 AFFECTIVE FORCES AND THE PRODUCTION OF IMAGES

By virtue of these attitudes, images have begun to disappear in the works of new art; the index of the representation of objects, faces, the illustrations of ideologies, the reflections of daily life, has dissolved and a new task has been put forth: that of expressing the sensations of forces that develop in the psychophysiological domains of human existence.

—Kazimir Malevich

4.3.1

For Bergson the image is not produced by the impression of light on the brain. This phenomenon—which is never an impression but always already a contraction and therefore a passive synthesis operated by a protosubjectivity—is rather the nascent movement of raw perception. The image, however, is constructed through a dynamic relationship between this raw perception and a schema, which is itself dynamic.

The relationship between the dynamic schema and raw perception is what Bergson calls intellectual labor or the labor of synthesis. According to him, besides the image, it is necessary to presuppose another type of representation that is capable of being realized in the form of an image but that, at the same time, is always distinguished from it. *The dynamic schema is therefore a mode of image production that is distinct from image representation.* Reducing all representation and all intellectual labor to a set of solid images, copied from the model of external objects is, according to Bergson, a very questionable hypothesis. The dynamic schema works not on images but on the force, the power, of images. This is another great Bergsonian hypothesis.

The dynamic schema is itself an intensive force, a force that works, in turn, not on the external form of the image but on its power. In some quite amazing passages, Bergson says that the dynamic schema does not merely connect to the visual or motor aspect of images, but mainly to their temporal relationships and powers. The image is therefore not primarily defined by its visual qualities, but by the relationships, especially temporal ones, between forces. Imagining only one relationship between actual and virtual images, as has been done previously, is an oversimplification that corresponds to imagining an attenuated set of relationships in the reconstitution of intellectual activity.

The solid image, copied from the object, is only the end result, the immobilization, of the relationships between powers that are recorded by our understanding. An image is the stopping of movement, its conclusion. The movement from which it is stopped is the movement of thought. The image is the end result of the work between powers, and this work should be understood as a real struggle between forces. This struggle and immobilization are easier to understand if we admit that the work of image production is not a representation, but a movement of representation. The image, according to Bergson, is a power, a relationship between hostile or allied powers.

4.3.2

To understand the movement of representation, we should reintroduce *affective force*, which as I have previously described, emergences and develops within the gap determined by the body in pure perception. We should reintroduce this force, otherwise the description of the production of mental images would be restricted to a mechanical one. This happens, for example, in the case of Edmond Couchot, who describes the transition from the analog to the digital paradigm as an overcoming of the optical model, but without departing from its mechanical logic. The stakes are high, since digital and simulation technologies reassemble these forces in a way that extends and transforms the human faculties—memory, imagination, intellectual labor—and the processes of the constitution of subjectivity.

The force that acts in the relationship between the dynamic schema and the image is attention, intellectual effort, or perhaps mental effort. Attention, for Bergson, is not the mere "sensory attention" that accompanies a simple perception, but a "tensive effort of the soul," an "acute desire." The intellectual effort that defines attention could be called the conatus of the intellectual activity that creates images and representations, which is sui generis power. Bergson distinguishes his concept of attention from individual attention, whose intensity, direction, and duration varies according to each individual. Bergsonian attention is a "mutual penetration," an intensive multiplicity. This brings us back to the concept of the virtual. The operation of intellectual effort, identical to the operation Bergson uses to define life, consists of a gradual transition from less to more realized—from intensive to extensive—of a reciprocal implication of parts in their juxtaposition. Attention thus is a virtual and affective force, while intellectual effort is the tension of this force that is actualized in distinct images and representations. Therefore, Bergson explains the intellectual labor of the creation of images and representations on the basis of the relationship between virtual and actual; that is, on the basis of a relationship between the active and passive dimensions of the force of time.

Gabriel Tarde, who developed a sociology from the same concept of force, and who actively participated in the debate that ensued in France between the late nineteenth and early twentieth centuries on the question of the relationship between attention, intellectual effort, and the production of internal states of consciousness, refers directly to attention as "pure" effort, pure tension, and to desire and its force. "So without attention, no sensation.... But what is attention? We can say that it is an effort to clarify a nascent sensation. But we must emphasize that effort, in its pure psychological aspect and abstracted from all concomitant muscular action, is a desire."[4]

The affective force that attention expresses is understood as an active force in the production of images and representation. The discussion on attention that developed in France at the turn of the last century has again become relevant with the rise of the information economy, which in the United States has been called the attention economy. Here we find some clues to understanding the information economy as an economy of affective forces.

But let us return to Bergson in order to pinpoint his definition of intellectual labor, or effort: "When we call to mind past deeds, interpret present actions, understand a discourse, follow someone's train of thought, attend to our own thinking, whenever, in fact, our mind is occupied with a complex system of ideas."[5] I have previously outlined a description of the work of memory as analyzed in Bergson's *Matter and Memory*. In a collection published as *Mind-Energy*, Bergson deepens the study of intellectual effort and offers one way to clarify these issues. The importance of the concept of intellectual labor or intellectual effort for my argument requires a precise reconstruction of the nature of intellectual activity in order to compare it with the work of technologies of simulation.

Let us summarize the analysis of *Matter and Memory*. Within memory, Bergson distinguishes different "planes of consciousness." These planes are infinite, but they are all located between pure memory, which is not yet translated into distinct images, and the same actualized memory of nascent sensations and movements. The work of recollection, but also the activity of interpretation and thought, crosses these different planes of consciousness. Intellectual effort consists in traversing them according to a specific intensity and direction. Until now, I have assumed a simple relationship between pure and actualized memory as a relationship exclusively between images. In fact, more profoundly, intellectual effort anticipates a relationship between the dynamic schema and images, which will form the heart of the following analysis.

To simplify the study of intellectual labor, Bergson presents different typologies ranging from the simplest activity, the activity of reproduction, to the more complex activity of production or invention. The simplest activity is the effort of memory or, more precisely, recollection. Bergson begins by distinguishing two types of recollection: automatic recollection and recollection accompanied by effort. The first is a form of automatic recognition, because the mind moves on only one plane of consciousness, while the second moves from one plane to another on all the planes of time that coexist within memory. The first is an instantaneous recollection, whose artifice consists in "making the mind move as much as possible among images of sounds or articulations without the more abstract elements, external to the plane of sensations and movements, intervening" (Bergson, *Mind-Energy*, 154).

Recollection accompanied by an intellectual effort makes us penetrate, on the contrary, into the world of the mind; that is, the world of forces. Memory does not consist in the mere capacity to retain auditory and visual images as a camera or phonograph does, but in a greater capacity to subdivide, coordinate, and link ideas (157). Recollection with effort does not consist in remembering an image, but in constructing a dynamic schema that operates in this direction. Memory presupposes a mechanism that cannot be simply reduced to the relationship between images. It is therefore necessary to have a representation that is not an imaged representation: one that constructs images, but at the same time differs from them. "I mean by this, that the representation does not contain the images themselves so much as the indication of what we must do to reconstruct them" (157; translation modified).

The dynamic schema, which is this very indication, is thus not an "extract of images," nor does it correspond more to the "abstract representation" of what the entire set of images means. The definition of this schema is difficult to establish, but by looking at studies into "the memory of chess players" who are "able to play several games at once without looking at the chessboards," Bergson suggests that the dynamic schema involves "force" and "power" and that images themselves are intensities, forces, and powers (158–59). Therefore, the relationship between the dynamic schema and images is not primarily a relationship between images, but between forces, between powers. Indeed, in Bergson's interpretation of the recollection effort of chess players, a mental vision of the pieces themselves is "more disturbing to them than useful. What they keep in mind is not the external aspect of each piece, but its power, its bearing and its value, in fact its function" (159). This description of the work of memory as the work of force and power is particularly striking: the bishop is an "oblique force," the rook is a "certain power of going in a straight line" (159).[6]

"What is present to the mind of the player is a composition of forces, or rather a relation between allied or hostile powers" (159). The player therefore obtains a representation of the whole that allows him to visualize the elements. It gives him "an impression *sui generis*" (159). I emphasize once again that this impression does not consist of visual or auditory images, or even readymade motor images. The impression represents itself "especially as indicating a certain direction of effort to follow" (161). It represents itself as a force or power.

"In these examples, the effort of memory appears to have as its essence the evolving of a schema, if not simple at least concentrated, into an image with distinct elements more or less independent of one another" (162). But we must distinguish between the different levels through which this dynamic schema operates: the development of the schema in images is immediate when there is a single image that presents itself to cover raw perception; but in general, several images are in front of raw perception. In the latter case, there is a process of coming and going, oscillation, struggle, and negotiation between the schema and images. "Hence a gradual modification of the schema—a modification required by the very images which the scheme has aroused and which may yet indeed have to be transformed or even to disappear in their turn" (177). The conclusion of the analysis of recollection effort is as follows: "The effort of recall consists in converting a schematic idea, whose elements interpenetrate, into an imaged idea, the parts of which are juxtaposed" (163).

Let us now turn to the highest form of intellectual effort: the effort of invention. The effort of invention is also constituted— like recollection and intellection—in the relationship between the dynamic schema and images. We start from a whole, from a schema that presents indistinct elements, in order to convert it into images and distinct elements. The schema is not a structure with immobile and rigid forms but is instead elastic and moving. Images are not solid, not fully defined and completed. Thus, there is a reciprocal relationship between the schema and images. The mind refuses to delineate precisely the contours of the schema, because in anticipation of its definition by the images themselves, they are captured by the schema to which they give consistency.

Intellectual labor consists in leading a single representation across different planes of consciousness from the abstract to the concrete, from the schema to the image. Let us follow the example, reconstructed by Bergson, of the effort involved in learning how to dance. All the elements mentioned above can be seen in this example, but in a new light. We begin by looking at the movement of the dance and have a perception of it. This perception, as we know, relates to memory and to motor images that, in this case, allow us to walk. But this raw perception is neither an arrested visual image nor even an image that is entirely visual:

> The image which we are going to use is not, then, a clean-cut visual image; it is not clean cut, because it is to vary and grow precise in the course of the learning which it is its business to direct; neither is it entirely visual, because, if it becomes perfected in the course of the learning—that is to say, in the course of our acquiring the appropriate motor images—the reason is that these motor images, called up by the visual image, but more precise than the visual image, invade it and gradually take its place. In fact, the useful part of the image is neither purely visual nor purely motor; it is both at once. (174–75)

Deepening his reflection, Bergson asserts that it is rather a question of outlining "the *relations*, especially temporal, between the successive parts of the movement to be executed" (175).

Rather than images—arrested and visual—these are temporal relationships. The dynamic schema is a sui generis representation in which relationships, forces, and powers are depicted. Rather than imagining representations, we must imagine movements of representations. Whereas the image is an arrest of movement, we recall movement in its duration, in its continuity and flow. The image is discontinuous, but memory is continuity, duration. "The schema is tentatively what the image is decisively. It presents in terms of *becoming*, dynamically, what the images give us statically as *already made*" (183).

The schema and images inspire a labor of mutual adjustment that is a veritable struggle, interference, and opposition. The schema will, as it continues, be filled with all the motor sensations corresponding to the executed movement and therefore construct itself. At the end of this process of action and reaction, we have an isolated, clear, and distinct image. We will be able to dance on the day the proposed schema obtains from the body the successive movements whose model it suggests.

> Thus, in the relatively simple intellectual effort in which consists the attention given to a perception, it seems indeed, as I said, that the pure perception begins by suggesting a hypothesis intended to interpret it, and that this scheme then draws to it manifold memories which it tries on the various parts of the perception itself. The perception, then, enriches itself with all the details evoked by the memory of images, whilst it remains distinguished from all other perceptions by the one unchanged label, so to say, which the scheme has affixed to it from the very beginning. (180–81)

4.3.3

To conclude this analysis of the concept of intellectual labor, I would like to show how we can locate here the actual-virtual circuit and how the relationship between active and passive forces—of which the actual and the virtual are the expression—alone can explain the production of images and, more generally, the creativity of intellectual activity. The relationship between the actual and the virtual can, in turn, account for the efficiency of machines that crystallize time, which only simulate, within precise limits, the actual-virtual circuit.

Bergson seems to have established, through the description of the relationship between the dynamic schema and images, an effective model of the operation of memory and intellectual activity, both in terms of interpretation and understanding as well as in terms of creation and innovation. Once this was achieved, Bergson pushed his questioning even further. Intellectual effort develops between a unique and invariable schema and a multiplicity of images that aspire to fill it. Within this model there may be a misunderstanding about the nature of the unity of the schema and the nature of the multiplicity of images.[7] "Will it be said that I am postulating the duality of *schema* and *image*, and also an *action* of one of these elements on the other?" (Bergson, *Mind-Energy*, 183).

Will there be, on one side, the unity of the schema as the unity of the mind and, on the other, the multiplicity of images as the multiplicity of the sensible? Will we fall back into the very opposition between the sensible and the intelligible that Bergson wants to avoid? In his explanation of the effectiveness of intellectual effort, Bergson begins by excluding the possibility of seeking difference outside of representation. Neither the psychological or affective accompaniments of the intellect nor any other force external to the intellect can explain it. "When the time comes to give an account of the efficacy, it will be necessary to leave out everything which is not idea, place oneself confronting the idea itself, and look for an internal difference between the purely passive idea and the same idea accompanied by effort" (182).

The internal difference makes it possible to explain intellectual effort without leaving behind attention and these active and passive forces. Intellectual effort is composed, as we know, of a mode of image representation and a mode that differs in nature. How to understand this composition, this "texture that differs from the inside"? To identify this difference, Bergson returns to the definition of the actual-virtual circuit. The relationship between schema and images proceeds from the difference between the actual and the virtual. Any other interpretation would lead us into a series of oppositions such as mind and matter, sensible and intelligible, which would merely reproduce the contradictions of modern philosophy.

The relationship between schema and image is a relationship between the virtual and the actual as forces, between what is in process and what is complete, between subjective and objective, between activity and passivity. Throughout his analysis of attention, Bergson describes the relationship between the dynamic schema and the image with concepts that define the relationship between the virtual—"indistinct elements," "different elements that interpenetrate"—and the actual: "distinct elements," "elements whose parts juxtapose."

If intellectual labor functioned only with arrested images, there could never be any creation or novelty, because these images "with their arrested contours" always represent what is actual and never power, the virtual, the subjective. We must therefore presuppose that "there must, besides the image, be an idea of a different kind, always capable of being realized into images, but always distinct from them. The scheme is nothing else" (184). The schema is of the order of force, of power, of the subjective, while the image is of the order of the objective, of completion, of passivity. Conversely, if images constituted the whole of our mental life—that is to say, if there were only passivity, objectivity—how could we define the creativity of intellectual labor? Faced with this difficulty, associationism assumes that images combine according to their resemblance. But in the case of intellectual effort, images combine not according to the laws of association, but rather according to an internal necessity, according to the principle of an internal force of differentiation. "The problem itself, therefore, must be standing before the mind, not at all as an image. Were it itself an image, it would evoke images resembling it and resembling one another. But since its task is, on the contrary, to call up and group images according to their power of solving the difficulty, it must consider this power of the images and not their external and apparent form. It is therefore a mode of presentation distinct from the imaged presentation, although it can only be defined in relation to mental imagery" (184–85).

Force and image: this is the relationship that can explain the creativity of intellectual labor. We must consider not only the image, but also its virtuality, its power, without which the mode of representation would be reduced to a grand tautology that explains nothing. "It is indeed as a function of real or possible images that the mental schema, such as it has appeared throughout this essay, should be defined.... The schema is tentatively what the image is decisively. It presents in terms of *becoming*, dynamically, what the images give us statically as *already made*" (183).

The schema and the image, the virtual and the actual, are forces distinguished according to spontaneity and receptivity.

But it is indeed the same force that splits itself each moment into activity and passivity, subjectivity and objectivity, virtual and actual. And it is in this way that we can account for the capacity of autoaffectation of the force of time discussed earlier, since the same force splits every moment into action and passion. Here, with the production of the mental image, we have an example of the functioning of subjectivity as a force of time.

One final remark: in the analysis of intellectual labor—and therefore of power—the relationship between cause and effect is largely inoperative. The gradual transition from "less realized" to "more realized" cannot be interpreted through the categories with which philosophy has attempted to capture the transition from power to action. The transition from intensive to extensive, virtual to actual, cannot be captured by the categories of efficient final causes. Intellectual effort is a process that escapes this conceptualization. "In analysing it, I have pressed as far as I could on the simplest and at the same time the most abstract example, the growing materialization of the immaterial which is characteristic of vital activity" (186). It is difficult even today, despite the rise of neuroscience, to find a more materialistic description of the activity of the mind based upon the internal differentiation of force according to spontaneity and receptivity, activity and passivity.

4.4 ON THE CONCEPT OF THE VIRTUAL

4.4.1

Based on this discussion regarding the relationship between the dynamic schema and the image, it is necessary to remove the ambiguities and misunderstandings that the term *virtual image* might introduce. Indeed, we cannot reduce Bergson's concept of the virtual to the concept of the virtual used to account for the production of simulation technologies.[8] The Bergsonian virtual image and the virtual image of digital technologies do not at all cover the same semantic field. It is clear that the production process of the virtual image by simulation technologies is closer to the possible-real relationship than to the actual-virtual circuit, according to the opposition defined by Bergson.

The virtual image produced by a computer is not, in the Bergsonian sense, a true creation, since its possibility is contained in the computer program—which has nothing of the Bergsonian virtual[9]—and not in its process of actualization.[10] It is a process of manufacturing rather than of creation. Nevertheless, in the production of virtual images, simulation technologies imitate the work of synthesis of intellectual labor. And just as video made accessible to us something of pure perception, simulation technologies make accessible something of the work of attention, something of that taking place that defines consciousness.

How to resolve this contradiction, since simulation technologies are indeed not virtual yet imitate it? As I have already indicated, the history of humanity and the evolution of nature could be described as the production of a panoply of "machines that triumph over mechanism" and of automatisms that release the possibility of choice. The affective force of time needs to be deployed by machines that release consciousness from the completion of teleological action in which it may remain trapped. These machines increase the capacity to insert indetermination and the unpredictable into mind and matter. Electronic and digital technologies provide a specific possibility of choice and action, because they imitate intellectual labor and, in a certain way, liberate it.

First of all, they increase the power—which we already saw with video—to desolidify material flows and bring us closer to becoming. They show us something of matter as it is described by Bergson: "Matter, looked at as an undivided whole, must be a flux rather than a thing."[11] Not to perceive "solids" and "states" as the intellect does is one of the conditions for escaping the completion of teleological action. Deterritorialization restores mobility to that which was frozen; it undoes that which the intellect had crystallized. Because "there do not exist things made, but only things in the making, not states that remain fixed, but only states in process of change. Rest is never anything but apparent, or rather, relative." If as Bergson states, reality is mobility, then "all reality is, therefore, tendency, if we agree to call tendency a nascent change of direction."[12]

Second, within this movement of deterritorialization that makes all of reality a tendency, simulation technologies reproduce the work of the synthesis of subjectivity—the synthesis of intellectual effort—which video sketched only crudely by adding memory to an apparatus of perception. We have seen how the ontological actual-virtual circuit (consciousness) takes on an actuality that might be described as the psychological aspect of intellectual labor. This subjectivity is neither a thing nor a mind but an oscillation between the virtual and the actual.

Intellectual labor consists in driving or contracting a representation through different planes of consciousness from the abstract to the concrete, from the schema to images. The abstract is never a representation, a substance, a state, but a movement of representation. The concrete is a contraction-solidification of this duration, an arrest of this becoming. However, digital technologies, within their mechanical limits, imitate this process of contraction and oscillation. Obviously, they cannot synthesize temporal relations and duration, but only discrete numbers: zero, one. As in Bergson's dynamic schema, computer programs do not contain fixed and completed images, but rather the instructions and movements necessary for constructing them. As in Bergson's schema, simulation technologies construct images through a "non-pictorial representation" different from images: the programming language.

Simulation technologies contract and synthesize nonrepresentative (asignifing but also aniconic) elements in the same way that intellectual labor synthesizes time as a nonrepresentative, asignifing element. *Like intellectual effort, simulation technologies produce images by synthesis rather than by impression.* The production of images no longer captures perception but can become imagination.[13] The technological apparatus, as a product of the understanding, can only imitate the synthesis of time by the organization of discrete elements. The dynamic schema and the process of oscillation can only have a discontinuous form. But the organization of these discrete elements approaches the work of the imagination as a synthesis of time. It is only by imitating the process of the crystallization of time that these technologies can also reproduce sensation, intellect, movement.

Félix Guattari's assertion that machines that crystallize time are technological apparatuses that imitate—in a hyperdeveloped and hyperconcentrated way—certain aspects of human subjectivity should be interpreted as pointing to the capacity of these apparatuses to reproduce aspects of the actual-virtual circuit that we have seen being articulated between the dynamic schema and the image. It is as if the intellect, in order to increase our capacity for action, was led to reverse its natural direction and provide us with the tools to think "true continuity, real mobility" and therefore to access creation. Even if technological apparatuses cannot attain the time of duration—"in which the past, always moving on, is swelling unceasingly with a present that is absolutely new"[14]—they can liberate duration and the time of mechanisms that neutralize it.

Nature and humanity have produced a myriad of machines—the brain, memory, language, concepts, society, and so on—that triumph over mechanisms, but only with electronic and digital technologies is it possible to become liberated from the fixed and completed image and to access the movements of representation. These technologies do not introduce a possibility of choice within the mechanisms of spatialization, but within those that fix time. Thus the term *synthetic image* corresponds more adequately to the definition of those images that are qualified as virtual. Synthesis is also the term used by Bergson to describe the action of intellectual labor as an intensive movement of consciousness.

4.4.2

Much has been said about digital technologies. My aim is simply to highlight one of the many aspects of these technologies; namely, that their power and productivity lie in the capacity to simulate and reproduce the relationship between the actual and the virtual with regard to intellectual labor or, as Bergson himself has described it, the work of synthesis.

The description of the transition from the video image to the synthetic image by artists is very close to the Bergsonian explanation of the transition from the sensory-motor image to the virtual image; that is, the transition from perception to memory, from the space in which images are lived and played to the space in which images are represented. Bill Viola, for example, says that once we are able to produce images without recording light, we are in the realm of conceptual space. This does not mean that we can reduce this space to the intellect and language. This aspect must be emphasized, in my opinion. In fact, Viola maintains a Bergsonian perspective, except that with Bergson the brain is not the seat of intellectual labor, the brain being a necessary (mechanical) but insufficient condition. "The real nature of a situation is not the visual image, but the information model of objects and space that the brain creates from visual impressions. The image is just the source, the input. Now, if you're talking about creating entire images that don't rely on light anymore, about building images from the point of view of conceptual space, then the mapping aspect comes out again."[15]

What interests me here, first of all, is Viola's idea that the visual impression does not truly characterize our relationship to the real. Second, for him the image tends to resemble a diagram. This concept recalls Bergson's definition of the schema. "The image [in 'Oriental culture'] is not considered to be a frozen moment or an arrested action or an effect of light or anything like that. . . . The image itself becomes related more to a diagram. A mandala, for example, is really a diagrammatic or schematic representation of a larger system, not necessarily the depiction of how an object appears to the eye."[16] The image is axiomatic, as we have seen with Bergson. Viola concludes by stating that to grasp these new forms of image production, we must abandon the models of the eye and ear and redirect ourselves toward the models of the processes of thought, of the conceptual structures of the brain.

The description of the operation of the synthetic image as encountered in the work of researchers such as Edmond Couchot corresponds to the Bergsonian description of intellectual labor, with the exception of the dynamic of forces. It is interesting to note that all the functions and mechanisms that Couchot describes, as the specificity of these virtual technologies, exactly overlap the relationship between the dynamic schema and the image I have just reconstructed. It should be added that programming languages, in their relationship with the image, are only a means to reproduce the power and forces of the actual-virtual circuit proper to the mind. The relationship between the matrix-image and the programming language seems to overlap with the Bergsonian analysis of the relationship between the apparatuses of nonpictorial representation and those of image representation. "The symbolic instructions that are entered into a computer have no sense, in the sense of the image.... The program language therefore remains outside the image and its figurative sense... but the language that underlies it remains inaccessible to the viewer, all the more hidden since it is not of the same nature as the image."[17]

The novelty that simulation technologies introduce (according to Couchot, though it is a widely shared view) is indeed a reproduction of the operation of intellectual labor. These technologies simulate and create images ex nihilo in the same way as Bergson's work of synthesis. And Couchot concludes in a Bergsonian manner: the programming language (the dynamic schema) functions as a sort of technological relay between the artist's intentions and the image. "In this, it is similar to the tool. But this relay is not material, it is symbolic, or at least both at the same time. It does more than translate the intentions of the creator; it helps in thinking and imagining. The computer frees the hand, but also thought, by automating new functions and forcing it to operate differently."[18]

Couchot establishes a radical difference between the production of synthetic images and the operation of human imagination. With Bergson, I have displaced this bias by reconstructing natural perception and imagination in a completely different way. An inadequate understanding of image production with natural perception leads to an ambiguous perspective on the technological phylum of apparatuses for image production. Jean-Louis Weissberg is struck by the overlap between the description of the process of vision in *Matter and Memory*—a process that is actively constructed by the relationship between the actual image and the virtual image—and the recomposition of vision carried out by computer-aided design. Since this comparison between the process described by Bergson and the operation of computers seems correct, it should be integrated into the description of the dynamic schema to develop its full heuristic potential. "This is the relationship between design and memory, such as computer aided design. The images are always partial extractions of a global model, itself invisible, because it is present with any angle, any lighting, any cut. The game consists in cutting, assembling, and linking the actual images displayed on the screen with the virtual images condensed and stored within the numerical model."[19]

4.4.3

The recording of images and sounds, and of sensations more generally, by digital technologies—unlike the production of images ex nihilo—is nothing other than a specialized development of functions already found in video, in particular the capacity to contract time-matter. The difference is merely the technological interface that crystallizes the time and duration of phenomena. The relationship between durations—the duration of time-matter, of pure perception, of technology—is controlled more accurately. Digital recording and reproduction also crystallize pure vibrations and shocks. Specific to digital technology is its capacity to more accurately distinguish the vibrations that compose all phenomena. Even more than video technology, it has the capacity to divide the waves, perturbations, and intensities that characterize images, sounds, and perception. This capacity to more accurately gather the infinite variations of matter makes it possible to reproduce them more faithfully. The temporality of the computer makes it possible to crystallize the passive syntheses that constitute us and the entire universe.

The enormous number of operations that the computer must execute requires a specific time and rhythm. The time of analog clocks, set according to astronomical time, is insufficient to perform all these operations. The computer in effect functions on a digital time generated by the fast and stable oscillations of an electrically excited quartz crystal. The characteristic of this temporality is that it can divide the seconds of analog time by a factor that escapes human representation. The computer clock thus enumerates extremely close instants, separated only by a few nanoseconds, with extreme accuracy and regularity due to the pulsation of the quartz.

The oscillations of quartz are still temporal scansions, separated by an interval, even if this interval is as small as possible. In any case, this time is always a spatialization of duration and not duration itself. But this spatialization of duration down to the nanosecond simulates, more accurately than analog time, the vibrations of pure perception. The frequency of this rhythm is so great that the comparison with a wave—even if it is imagined to be composed of the most discrete elements—becomes possible. Digital technology makes it possible to follow and crystallize all the wave phenomena of which pure perception is composed.

The computer can more accurately reproduce any visual or sound phenomenon because it introduces two new conditions:

discretization and calculation. Discretization is the ability to transform a temporal phenomenon into a spatial one. It can take snapshots and cut discrete elements into becoming (duration). But the break in the continuity of the becoming of phenomena is so infinitesimal that it offers us an illusion closer to reality, closer to what has been described as the "image in itself." "Discretization is the substitution of a signal, for example, made of an infinite number of consecutive values, which can be infinitely close to one another, by a finite number of values, which can only be taken in a finite number of possibilities."[20]

5

NIETZSCHE AND TECHNOLOGIES OF SIMULATION

5.1 SIGN AND FORCE, REALITY AND APPEARANCE

5.1.1

How might virtual technologies be defined as Nietzschean? What does Friedrich Nietzsche show us? He demonstrates that humans operate through simulation; that we utilize nonexistent things such as points, lines, or atoms; and that we require semiotics, images, the fiction of the subject and object, and concepts of causality in order to think and live. We need to divide time and space. We affirm form, Nietzsche suggests, because we do not grasp the subtlety of absolute movement; we acknowledge identity and permanence because we can only *see* the permanent and can only remember that which is similar or identical. According to Nietzsche, all these concepts are simulations, errors, fictions that we use to determine a certain regularity and stability over time, enabling us to desire. What difference is there between the use of a computer language (zeros and ones) and the many methods humans implement to make the world knowable? For Nietzsche there is no difference, since the degree of falsity is measured not on the basis of a real or true world, but rather by the power it is able to accumulate.

Humans have always worked with and through simulation. Virtual technologies only render certain qualities of this particular animal reproducible and automatic. We find that the simulation function, previously limited to the brain and nervous system (or more precisely, to the specific assemblages of the forces called "man," since locating these functions in the brain is difficult), instead of diminishing, extends and amplifies through automatic calculation. Not only does simulation extend and amplify, it becomes a machinic assemblage. Today this force of simulation is assisted by the computer, and the machinery of simulation frees the computer from all its inert schemas. Instead of operating in the world through simulations such as the point, the line, the subject, the object, and so on, we can now operate with far more sophisticated simulations. The old forms of simulation were just as "natural" as the new ones. And just like the new ones, they function as an interface between signifying and asignifying flows, according to a specific will to power that no longer belongs to "man."

But do the new forms of simulation make the real disappear? For Nietzsche, this question is completely absurd since it masks a will to power that it does not want to recognize. On the contrary, these new forms multiply the degrees of reality. The multiplication of the strata of reality entails a redefinition of what we have called the real, because these technological assemblages rearrange it in the infinitely small and the infinitely large. So what are these new technologies? They are fictions that allow us to act on new levels of the real. With the field of quantum physics and the new social machines, language and productivity can no longer be qualified as human. They are simply fictions that allow us to understand and act beyond the human and its world.

5.1.2

Nietzsche's philosophy is constructed on the relationship between affective forces and signs. Increasing the capacity to sense and act proportionally increases the capacity to construct forms and invent signs. According to Nietzsche, forms and signs should be interpreted as a language through which forces can be apprehended.

"What's essential is the evolution of forms that represent many movements, the invention of signs for whole species of signs."[1] For Nietzsche, a multitude of movements exist within and outside us, of which we are completely unaware. The only way for us to come to know them is through semiotics. "I deny that these movements are created by our will; they are carried out within us but remain unknown to us. We can grasp their unfolding only in symbols (tactile, colored hearing) and in some fragmented moments."[2]

Our logic and our sense of time and space are "faculties of incredible abbreviations" (accelerations) that constitute our ultimate strength, even though they reduce the experience of signs and the huge mass of things that can be grasped. Or to use the language of Félix Guattari,[3] the enrichment of semiotics is one of the preferred methods through which capitalism is expressed (the semiotization of the faculties that constitute subjectivity, of stratifications that compose matter, of the complexity of social stratifications, etc.). Capitalism thus develops technological assemblages that not only reproduce our capacity for semiotization but also deploy this force beyond the abilities of our senses or their faculties of simulation. This is certainly, for Nietzsche, a symptom of the increasing capacity to sense (to affect and be affected) and is therefore the condition for an increase of power. The fact that semiotization increasingly pushes the boundaries of the real does not lead to a disappearance of the world, as if it were overcome by the appearance of the sign. This is just a symptom of the capacity to take control of an enormous mass of facts with signs. The sign does not refer to a structure or code, as in academic semiotics, nor does it refer to intersubjective agreements, nor to generic society and humanity. The sign refers back to neither the sign, reality, nor convention, but directly to power, since it is a symptom of the will to control and create.

"It is the powerful who made the names of things into law and, among the powerful, it is the greatest artists in abstraction who created the categories" (Nietzsche, *La volonté de puissance*, 57). The source of signs lies in the power of control and creation, not in a relationship between truth and the world. There is no true or false, only a relationship between the abbreviation of signs and signs themselves. The fundamental thought underlying this position is the following: the outside world as designed by our semiotics is nothing but a sum of value judgments. It is difficult for semiotics (which in one way or another always revolve around the relation between denotation and connotation)—but also for the tradition of analytic philosophy (see, for instance, how Ludwig Wittgenstein deals with the apprehension of colors)—to understand this affirmation by Nietzsche: "Green, blue, red, hard, soft are inherited valuations and their signs" (248).

5.1.3

For a semiotics to be possible, it is necessary to introduce identity and analogy and, with them, a simplification. Absolute differentiation would be impossible to pin down in its perpetual metamorphosis. "If I believe that the subject could emerge from the illusion of identity, for example when various forces (light, electricity, pressure) produce an identical stimulation on a protoplasm, I would conclude that there is an identity of causes" (238).

Thus, according to Nietzsche, *belief (judgment) must precede the subject* and self-awareness. The key thing is that inaccuracy, imprecision, and error of judgment lead to a certain simplification of the world. This belief is already inherent in the simple phenomenon of the assimilation of the protoplasm. If there are truths, they reside in the inorganic world. Error and appearance begin with the organic world. Does simulation make us lose our humanity? On the contrary, says Nietzsche, it is a quality of the living. "An increase in simulation is proportionate to the rising order of the rank of creatures. It seems to be lacking in the inorganic world . . . cunning begins in the organic world . . . a thousandfold craftiness belongs to the essence of the enhancement of man. Problem of the actor" (313).

If the hierarchy of beings is measured by their capacity for simulation, could machinic development signal an increase in the hierarchy rather than a degradation? Or at least, does it not set the conditions for this increase? For Nietzsche, this situation is certainly a symptom of an increase of power.

5.1.4

"All is spectacle, the world is directed by technologies of vision, we have lost the real!" cry the beautiful souls. But this is not true, Nietzsche suggests, because if we turn appearance into a duplication, an image, a simulation of a real world, that world would be here, hidden behind or beyond its double. But if we abandon this simulation, appearance ceases to be the opposite of being and would no longer be a mask. "Appearance itself belongs to reality, it is a form of its being. . . . Appearance is an arranged and simplified world in which our practical instincts have been at work" (329).

Appearance and being are posited simultaneously, in one and the same movement. There is not a structured world on the one side and a perception that seizes it on the other. Appearance is constitutive of being. And appearance, before being a condition of perception, is a condition of activity, of power. Appearance is therefore an instrument for the increase of power, the increase of power to affect and be affected. From this perspective, virtual technologies can be considered more "true," since in denaturalizing perception, they reveal the operations of its apparatus. With the advent of virtual technologies, the image leaves the screen and becomes a "place" where we can act: meet someone, have sensory experiences, and so on. We are no longer just in front of the image; we can also be in it: the boundary between spectator and actor becomes increasingly blurred. The image is no longer a copy, a trace, or the presence of an absence, but a full, positive immediacy. The virtual image thus allows us to understand the following Nietzschean affirmation practically: "What is appearance to me now?. . . To me, appearance is the active and living itself."[4]

Instead of subtracting us from the real, appearance conceals the mechanisms of its constitution. The fear that this new simulation might cause is probably due to the fact that it is an appearance that escapes our habitual thoughts and actions, being neither the real nor its copy, neither matter nor spirit.

5.1.5

We are told that the new technologies of vision and simulation are "a cage for the eye." But has it ever been otherwise? The gaze is a question of ethics, as Nietzsche reminds us; the new vision technologies simply make the question unavoidable. By rendering the image independent of light and suturing it instead to mathematical (nondiscursive) language, such technologies highlight their conventional character, to repeat one of Nam June Paik's statements. The Nietzschean problem of typologies resurfaces: who sees, who perceives? Namely, what apparatus of power, what form of subjectivation sees/perceives? *Because what we see is never the real* but instead the series of effects produced by fictions and errors organized in a certain way and under certain conditions by and for the human (subject, object, divisible time and space, etc.). We see more with the "will to truth" than with the physiology of the eye. According to Nietzsche, the will to imprint the nature of being with becoming is the highest will to power expressible by the human. "Here we are asked to think an eye which cannot be thought at all, an eye turned in no direction at all, an eye where the active and interpretative powers are to be suppressed, absent, but through which seeing still becomes a seeing-something, so it is an absurdity and non-concept of eye that is demanded."[5]

It is not just the brain's active involvement in perception as recognized by any theory of vision but, more profoundly, the fact that these active forces refer to the will to power. Nietzsche emphasizes, as we have already seen with Henri Bergson, that the optical model connected to the physiology of the eye explains nothing about vision, since for Nietzsche sensory perceptions are already acts. "For something to be perceived, it must already function as an active force that accommodates stimulation, makes it act, and adapts to it according to stimulation and modification."[6]

This active force is the original affective force (pathos), a duration that conserves the before in the after, death in life. Electronic technologies liberate us from the naturalistic illusion of vision and bring our attention to these active energies because, in reality, our perception is the result of a double forgery, one having its origin in the senses, the other in the mind, the purpose of which is to produce and maintain a world of being and of equivalence. But rather than conceiving a politics of the eye—it takes more force, more morality to see the unknown as the known—it is usually preferred to demonize technology. The virtual, instead of preventing us from seeing, is a pedagogy of vision and perception, of their mechanisms and affective forces. Once it is established that all vision is more or less virtual, the problem of the specific difference of computer-assisted vision arises.

5.1.6

The Nietzschean explanation of vision through mechanisms and forces also reminds us of Bergson's analysis. Vision is essentially produced by "active forces" that trace, in advance, a free image onto the stimuli that affect us. This free image is essentially different from the actions occurring in our bodies that remain inaccessible to us. The function of this image is comparable to the "dynamic schema" in Bergson's description of the process of vision. "This image is very general, it is a schema. We imagine that it is not only a guiding thread, but the impulsive force itself."[7]

For Nietzsche, the origin of this free image must be sought in poetic force, in the "divinatory instinct" that must divine, according to real elements, unknown things.[8] The force that explains the construction of the eye is thus an aesthetic instinct that idealizes the human and nature in a pictorial form. Perception is organized by aesthetic forces, but as with Bergson, it is always a subtraction and reduction from pure perception.[9] The simplification of the real is followed by the addition of forms.

"There is a faculty within us that makes us perceive the main features of the mirror-image with greater intensity, and another faculty that makes us beat to the same rhythm, even beyond the imprecision of the sounds. It must be a *force of art*, because it creates. *Its primary means is the omission and non-capture of much of what we see and feel*. It is therefore anti-scientific because it is not equally interested in all that is perceived."[10]

This artistic faculty has a dual function: it produces and selects images. Producing and selecting images are never voluntary acts but are directly related to imitation and result in the transposition of what is imitated into another register. Our senses, according to Nietzsche, imitate nature by falsifying it ever further. Identifying the similar with the similar, discovering resemblances between things, corresponds to the original process. "Memory relies on this activity and employs it continuously. The exchange of one thing for another is the original phenomenon."[11]

The imagination, understood as the quick perception of resemblance, functions according to the same principle. But what is the force that compels us to imitation and resemblance? The appropriation of an unknown impression by means of something known. Every perception is a multifaceted imitation of stimulations that transform it in different domains. This operation is common to all organic beings and occurs, except in the lowest evolutionary strata, by the production of images; it is already present in the malleable unconscious force that we see in reproduction. Therefore, sensory perceptions are not primary but, on the contrary, are constructed from that what Gilles Deleuze, following Bergson, calls passive syntheses. Sensory perception takes place without our knowledge, and everything we are conscious of is already a developed perception. "The same leveling and ordering force that reigns in the protoplasm, reigns in the phenomena in which we incorporate ourselves within the external world. Our sensory perceptions are already the result of this assimilation, this identification with our entire past. They do not immediately respond to impression."[12]

The optical model takes into account neither the role of aesthetic forces nor the role of the body and its work of simplification, assimilation, and falsification. As in Bergson, the entire model of perception, built upon the separation between the sensible and the intelligible, is put into question. The space of representation is, for Nietzsche, the only way in which the eye of a particular will to power apprehends what happens. It does not correspond to anything real. "Subject, object, an actor added to the act, the action separated from that which provokes it: let us not forget that this is mere semiotics and nothing real. Mechanistic theory as a theory of motion is already a translation into the sense language of man."[13]

This implies that the vision that corresponds to the model of representation is itself a relationship between flows: asignifying flows of intensities, the complexities of what happens, and the signifying flows of the semiotics of subject, object, and truth. "In a world that is essentially false, truthfulness would be an unnatural tendency: it could have meaning only as a means of attaining a higher power of falsehood."[14] The virtual image introduces us to the world of falsehood, the world of lies. Only our bias toward truth prevents us from feeling at ease. (Children, in whom the habit of truth has not yet been set in stone by language, have a completely different relationship with these technologies.)

5.1.7

When machines automate the functions of semiotization and simulation, they highlight the differential element, the active element—the will to power—that is, the affective force, the capacity to produce events.

Simulation technologies liberate us from mechanism and the organism. Not only do they introduce us to other dimensions (those of dynamic quantities, variations, and intensities), they simultaneously introduce us to the dimension of the event. From this perspective, it is fair to say that the automatic reproduction of the capacity for semiotization and simulation liberates the virtual as a space for action. What attitude should we assume regarding these technologies once they have denaturalized the subject, object, and image and have involved them in an increasingly thorough deterritorialization, releasing active forces? Technologies of simulation inspire us to assume, without any recourse to transcendence, that we are artists, inventors, experimenters. Any other attitude would simply reintroduce an apriority that imprisons these forces and subordinates them to a simulation that would be external to them.

> Just as little do we see a tree precisely and completely, with respect to leaves, branches, colors, and shape, we find it so much easier to imagine an approximate tree instead. Even in the middle of the strangest experiences we do the same thing: we invent most of the experience and can barely be made *not* to regard ourselves as the "inventor" of some process. What all this amounts to is: we are, from the bottom up and across the ages, *used to lying.* Or, to put the point more virtuously, more hypocritically and, in short, more pleasantly: people are much more artistic than they think.[15]

Nietzsche assumes that in most cases what our eyes see is not a sensory impression but a product of the imagination. The senses only provide us with pretexts, tenuous grounds that we further develop.

Virtual technologies remind us of Nietzsche's critique of the true and the real. They cause the ideal of truth, but also the world of appearances, to crumble. The world of technology is a false world. The pixel is beyond good and evil; produced by algorithms, it contains nothing real. Therefore, the only practice adequate to it is that of the experimenter, the artist. There is nothing to reproduce, but everything to create. The deterritorialization of spatiotemporal coordinates, the liberation of the flow of forces, and the emergence of being before the subject-object relationship constitute the material conditions for accessing the last stage of the will to power: becoming-artist, willing-artist. This willing-artist is the desire to create truth, because truth is "not something there, that might be found or discovered—but something that must be created and that gives a name to a process."[16]

There is no other truth than the creation of the new: not the creation of new forms, but of the powers of metamorphosis. The image matrix compels us to this becoming-artist and, in this sense, might be a powerful vector of subjectivation. Its power lies not in being a model of something, the presence of an absence, but in its capacity to transform. The willing-artist can thus be the form of subjectivation most adequate to this power, because it is a force that no longer judges life in the name of a higher, transcendent authority. The willing-artist evaluates all forces in relation to the life in which it is implicated. The willing-artist is a force that creates and transforms itself from the forces it encounters and composes with them an ever-increasing power, always opening up new possibilities, new forms of life, and new existential territories.

5.2 REAL TIME REDUX: SPEED AND THOUGHT

5.2.1

Now let us return to the problem of real time that I already discussed regarding video technology. The process of deterritorialization leads to a multiplication of temporalities and speeds. Real time is a term used by IT professionals to define the speed of certain computer operations.[17] However, according to the critics of real time, its absolute speed (the speed of light) would cause the loss of our notion of space, the simultaneity associated with this speed would cause the loss of our notion of alterity, and its temporality of the absolute present would cause the loss of experience. In short, the subject's form of perception and representation would be lost. Such a subject would no longer be able to act—which would entail the end of politics, the end of history—because consciousness and reason would no longer be the determinants by which action is calculated.

I have already demonstrated that this critique of real time is contained within a concept, thus depriving itself from seeing the plurality of temporalities and speeds that the new paradigm brings. It takes as exclusive the will of apparatuses of power that reduce the multiplicity of temporalities and speeds to the absolute speed of light. That real time is the time of the absolute present (instantaneous time that is reduced to an eternal present) or the time of the event (emerging from another block of space-time), and it never depends on technological machines but on apparatuses of power. The only way to talk about real time is to first install oneself within it and submit to its flow. If we do this, and abandon the fictions and errors of our natural perception, we will witness a remarkable spectacle: a veritable cybertrip. But then again, we should entrust our perception to Nietzsche's "third eye" rather than to cyberliterature. I could also draw upon the latest discoveries of cognitive science that seem to reach similar conclusions to those of this great visionary. But I still prefer Nietzsche, since cognitive science uses Nietzschean hammers to reconstruct a theory of equilibrium.

We will now return to the theme of time and analyze it in relation to thought and action, which have become scientific and—more importantly—productive and industrial issues. The information economy has pushed cybernetics to develop because, as a cognitive technology, it objectifies and makes almost physically palpable the relationship between brain and thought, between speed and knowledge. Cybernetics helps us understand how we think. Nietzsche's critique of the identification of thought with consciousness and language now seems to be echoed, albeit mutely and indirectly, by minor currents in cognitive science that came after the failure of the program of artificial intelligence based on the categories of representation, denotative language, and logical thinking. Indeed, computers and their simulations of thought and action impose upon us some Nietzschean truths: First, thought and action overflow consciousness and language from all sides. Second, thought and action are multiplicities; reducing them to consciousness and language is the surest way to destroy them.

5.2.2

Computer simulations externalize the role of consciousness and are subject to practical operations. If some of the functions attributed to consciousness can be automated, this does not mean that machines can replace us by becoming more human but only that, until now, consciousness and the human were assigned roles that did not correspond to reality. New technologies simply help us to put consciousness in its proper place. Nietzsche already had a very clear idea about the functions of consciousness, so its automation would have come as no surprise to him. "Consciousness in a subsidiary role, almost indifferent, superfluous, is perhaps destined to vanish and give way to a perfect automatism."[18] So would the automation of consciousness have positive outcomes? For Nietzsche, consciousness misleads us in many ways; for example, about the happenings and operations of our brain. This is due to, among other things, the brain's slowness, the fact that it cannot function in real time. "The suddenness with which many effects stand out misleads us; it is a suddenness only for us. There is an infinite number of processes that elude us in this second of suddenness. An intellect that saw cause and effect as a continuum, not, as we do, as arbitrary division and dismemberment—that saw the stream of the event—would reject the concept of cause and effect and deny all determinedness."[19]

Machines that crystallize time enable us to see and feel something of these flows, in which we find—just as we do within the brain—nonhuman speeds. These nonhuman speeds are multiplicities rather than mere contractions of time (its infinite acceleration). This second shock, as Nietzsche calls it, is not perceived as absolute speed, since it refers to the speed of consciousness, whose intensity depends upon the facility and velocity of cerebral transmission. What Nietzsche did conceptually, cognitive technologies have made tangible, putting time and speed into thought. Consciousness is a terrible way to approach this new dimension, this new anthropology, because "between two thoughts all kinds of affects play their game: but their motions are too fast, therefore we fail to recognize them, we deny them."[20] Not only does the brain operate at a speed that cannot be perceived by consciousness; this lack of real time misleads us about the logical and rational nature of our thoughts.

5.2.3

The failure of the real time of consciousness means that it only can see and record the end results of the thought processes that occur completely unconsciously. *This defect prevents it from recognizing the multiplicity that constitutes it.* It reduces the multiplicity of speeds, of human and nonhuman agents, of forces and wills, to a rational and linguistic form of thought. Artificial intelligence, which tried to follow this path, ended up at an impasse. Logic—the model of thought and action that served as a reference point for artificial intelligence—reduces the multiplicity of thought to a single element, a rational element, and subtracts all other forces that contribute to its constitution. The impasse that has determined the practical application of the simulation of human thought and action by automatons has shown that the rules of thinking and acting are not rational rules that correspond to the true relationships between sensibility and intelligibility, subject and object. If they were rational rules, they could be abstracted, explicated, and reproduced, enabling machines to behave properly. But this is not the case, even in situations as simple as moving an automaton through a space.

The logical model of the operation of thought and action is only adequate for situations in which the world is predefined and in which, consequentially, sense is presupposed and cognition is reduced to the mere manipulation of symbols.[21] When the value of the sense of a situation, a word, or a sign is defined, cognition can be reduced to a manipulation of discontinuous elements that come to represent, unequivocally, that to which they refer. A game of chess almost perfectly meets these conditions.²² But any other situation would involve insurmountable difficulties for the logic of artificial intelligence. The impasse of this project has forced researchers to assume that thought and action do not function according to logical models, but according to forces that cannot be defined as rational. Nietzsche can help us understand how our thought functions and how our actions are organized as well as to identify the aporias of cognitive science. According to him, thinking as defined by the theorists of knowledge is something that never happens. "We believe that thoughts as they succeed one another in our minds stand in some kind of causal relation: the logician especially, who actually speaks of nothing but instances which never occur in reality, has grown accustomed to the prejudice that thoughts *cause* thoughts."²³

However, our mental life follows from everything else. In the passage from one thought to another, a completely different world intervenes, such as the instinct of contradiction or domination. But more profoundly still, thought is essentially a multiplicity of thoughts: thinking through concepts, thinking through images and sounds, tactile thinking, but also organic and preorganic thinking. To understand this, we should first establish a relationship between percept and concept. Nietzsche ironically defines this relationship as one of filiation. If the image is not the mother of the concept, then at least it is the grandmother, because the eye, in vision, functions exactly as the mind understands. "First *images*—to explain how images arise in the spirit. Then *words*, applied to images. Finally *concepts*, possible only when there are words—the collecting together of many images in something nonvisible but audible (word). The tiny amount of emotion to which the 'word' gives rise, as we contemplate similar images for which *one* word exists—this weak emotion is the common element, the basis of the concept" (Nietzsche,

La volonté de puissance, 44). But the percept-concept relation does not exhaust the definition of thought, because for Nietzsche, it is beyond any doubt that we think with images and sounds but also with tactile sensations. "There must have been thinking long before there were eyes: 'lines and shapes' were thus not originally given. Instead, thinking has longest been based on the sense of touch. . . . Thus, before we started practicing our understanding of the world as moving shapes, there was a time when the world was 'grasped' as changing sensations of pressure of various degrees. There is no doubt that we can think in pictures, in sounds: but we can also think in sensations of pressure" (253).

But Nietzsche goes even further by arguing that there is thought even in the most elementary organic phenomena, because even the protoplasm can "judge." Under the influence of certain stimuli, it creates a simplified whole by eliminating numerous details. It poses an identity by affirming its own existence. If thinking consists of integrating new data into old patterns, then even crystals think. "'Thinking' in the primitive state (pre-organic) is the *effecting of shapes*, as in the crystal" (238).

The important thing to note is that all these forms of thought—thinking through concepts, images, sounds, touch, cells, the preorganic—coexist and multiply within the individual. This multiplicity is coordinated, hierarchized, but always present and constantly reproduces itself. Thinking is not the result of an evolution from preorganic thought to conceptual thought. Nor is it the end point of a dialectical surpassing of its origins. How does it operate then? In thought, there are illogical forces of a multiplicity that, far from complying with logic as a rule of conduct, engage in a merciless battle for control. "The course of logical thoughts and inferences in our brains today corresponds to a process and battle of drives that, taken separately, are all very illogical and unjust; we usually experience only the outcome of the battle: that is how quickly and covertly this ancient mechanism runs its course in us."[24]

Even the logical form is not necessarily logical, but corresponds to an imperative of power. Today cognitive psychology, aided by real simulations of intelligence technologies, recognizes the modularity, heterogeneity, and multiplicity of thought. It suggests that the mind is made up of many agents and therefore appears as a kind of "society." The connectionist approach—which is the basis of cognitive psychology—even seems to establish a relationship between signifying and asignifying flows in the determination of sense. Sense is no longer confined within discrete symbols, as it was in the rationalist trend, but becomes a function of the overall state of the system (the brain) and remains connected to the general activity in a particular area such as recognition or learning. This overall state emerges from a network of entities that some theorists have called subsymbolic. Sense and formal principles then find their origin in the intermediate state between the biological and the symbolic. However, cognitive psychology does not associate the multiplicity of this global state with battle or the will to power; it refuses the forces that act and feel. A multiple thought, yes, but rationalized and pacified. What cognitive psychology sanitizes, Nietzsche returns to us in its warlike form, bloody and seething.

> Before knowledge is possible, each of these impulses must first have presented its one-sided view of the thing or event; then comes the fight between these one-sided views, and occasionally out of it a mean, an appeasement, a concession to all three sides, a kind of justice and contract.... Since only the ultimate reconciliation scenes and final accounts of this long process rise to consciousness, we suppose that intelligence must be something conciliatory, just, and good, something essentially opposed to the instincts, when in

fact it is only a certain behavior of the drives towards one another. For the longest time, conscious thought was considered thought itself; only now does the truth dawn on us that by far the greatest part of our mind's activity proceeds unconscious and unfelt.[25]

Nietzsche proposes a definition of thought that evades the imperialism of consciousness and instead asserts that the most important problem for thought is the powerlessness to think its own outside. What is exciting about simulation technologies is that confining thought within itself is no longer a philosophical limit but immediately becomes an industrial limit, since it prevents the development of automatons.

In a way, cognitive science is forced to return to the path indicated by Nietzsche: thought is not a single and generalizable ensemble of logico-mathematical organization that operates in all domains. Cognitive psychology has shown that the building blocks of thought (hearing, speech, linguistic understanding) operate automatically and coordinate with each other without the intervention of conscious control. Therefore, it becomes possible to say that the technological automatisms of the computer "externalize" the physiological automatisms of the brain (its functional speeds and slownesses). But what do the technological automatisms liberate? They liberate one element of the will to power: the outside of thought, the unthought. It is this element that the sciences have much more difficulty recognizing.

5.2.4

Nietzsche's philosophy not only suggests that the strong program of artificial intelligence is in crisis because it assumes that axiomatic logic is adequate for the apprehension and mastery of reality and action. More profoundly, he asserts that the logical principle contains not a criterion for truth but an imperative with respect to what must be held to be true. In other words, logic is not primary but results from the will to power, the assemblage that it presupposes. Event and assemblage are subtracted from truth, from good and evil. The subject and object of knowledge, the subject and object of affect, the perceiver and the perceived, are all effects of the event, the assemblage, the imperative that grounds them.

"And so symbols, words, and rules that seem capable to account for themselves, to realize the establishment of relations of behavior for this purpose—which the failure of artificial intelligence, who took this appearance to be 'real', has shown—are simply order-words whose only value is their relative redundancy with an assemblage."[26] This is the imperative of which Nietzsche speaks. The only theory that comes close to these flashes of truth is the minor current of cognitive science. It accepts that the failure of research into artificial intelligence was its proposal to apply to the behavior of automatons the concept of action that has just been criticized. Francisco Varela—a specialist in cognitive science who has drawn out the most radical consequences of this failure of the strong program of artificial intelligence—has proposed the notion of enaction to describe the irreducibly contemporary emergence of a world, which takes both meaning and acting as ways of relating to the world. Action is not the submission to rules leading toward a goal in an already given world. Knowledge is not a representation of the world, its rules, its goals. In Varela's words, action and knowledge are ontological: "Perception does not consist in the recovery of a predetermined world, but in the perceptual comportment of action in a world inseparable from our sensory-motor capacities. In other words, cognition consists not in representation, but in embodied action. Correlatively, we can say that the world we know is not predetermined, but emerges according to the history of structural couplings."[27]

We can also detect the Nietzschean categories of multiplicity, speed, and the unconscious in Varela, albeit stripped of their power. "Moreover, the linchpin that articulates enaction . . . consists of a rapid, non-cognitive dynamic in which many alternative micro-worlds are active."[28] The conclusions correctly assess the critique of the concept of the subject as it is constituted in the modern philosophical tradition. Consider two aspects that Varela defines as post-Cartesian:

> 1) Knowledge is always a mode of knowing, constituted on a concrete basis. What we call the "general" and the "abstract" take part in the disposition of action. Knowledge is always "ontological;" it is not the knowledge of beings it would adequately represent, but rather the reciprocal production and specification between the knower and the known. Cognition is not the submission to rules. It is first of all a condition for all explicit or implicit rules, the creator that gives sense to these rules.
>
> 2) The micro-worlds are not coherent or integrated within some enormous totality that would determine even the smallest fragments. It is rather a kind of interaction, of turbulent conversation. It is the very presence of this turbulence that allows for the creation of a cognitive moment, with the specificity of a system and its history.[29]

Why do I insist on a theory put forth by the "English psychologists"—according to a formula used by Nietzsche to criticize the moralists—despite the presence of elements that seem akin to Nietzsche's critique of the faculty of thought? Because the element that establishes the relationship between knower and known, between perceiver and perceived, is identified by the autofoundational interaction between the elements involved. The hermeneutic circle here constitutes the more or less acknowledged reference. But the concept of representation is substituted by that of interpretation. Indeed, most thinkers who refer to this new method of artificial intelligence are part of the phenomenological tradition and European hermeneutics. "These thinkers are in fact concerned with the entire phenomenon of 'interpretation,' in the circular sense of the connection between action and knowledge, between knower and known. We refer to this complete circularity between action and interpretation with the term 'making-emerge.'"[30]

5.2.5

Technologies of simulation have rendered irrefutable the ideas that we think with and in time and that thinking is an event. This brings us back to the speed Nietzsche speaks of. We must abandon the subordination of time to the speeds of consciousness in order to live the time of the event, the "untimely." The technologies of time liberate us from inert time and force us to think of time as the unthought of thought and of action as its outside. The specificity of the temporality of these technologies goes beyond the connection established by Paul Virilio between speed and time. The infinite acceleration of time does not lead to its disappearance but opens to the time of the event, the time of creation, since this technological apparatus reveals another block of space-time.

The concept of time as nonchronological, as the time of the event, is the key that allows us to overturn the theoretical perspective as well as practices of communication. This is also how we can initiate a radical critique of the concept of action and its new surrogate, interaction. The introduction of real time by the new technologies has detonated the concept of action and replaced it with the event. Video technology was involved in this becoming. *By deterritorializing subsequent flows, digital technologies push and encourage us toward a knowledge, thought, and action of the event and situations, a knowledge and action of assemblages and multiplicity*—a knowledge and action in which consciousness is devalued, in which it is no longer the highest element of the human but rather a form of communication. It was always within the limits of our perception, starting from the impossibility of consciousness to see beyond divisible space and time—and therefore the impossibility of seeing other temporalities—that the concepts of action and actor have been constructed. According to Nietzsche, under the seductive influence of language, and the fundamental errors of reason that have sedimented within it, we understand all action as conditioned by an agent, a subject. We distinguish lightning from its flash and take the latter for action, for the effect caused by a subject called lightning. The mistake is to split action. Nietzsche asks us to refuse this splitting by moving from action to the event. There is nothing lying beneath the action, the effect, the becoming; action is everything.[31] "An artificial distinction is made in respect of events between that which acts and that toward which the act is directed (but this 'which' and this 'toward' are only posited in obedience to our metaphysical-logical dogmatism: they are not 'facts'). . . . We have not got away from the habit into which our senses and language seduce us. Subject, object, an actor added to the acting, the acting separated from that which it does: let us not forget that this is mere semiotics and nothing real."[32] This model of action is also the model from which artificial intelligence aims to construct the action of automatons. "The 'spirit,' something that thinks: where possible even 'absolute, pure spirit'—this conception is a second derivative of that false introspection which believes in 'thinking': first an act is imagined which simply does not occur, 'thinking,' and secondly a subject-substratum in which every act of thinking, and nothing else, has its origin: that is to say, both the act and the actor are simulations."[33]

5.2.6

Perhaps the most important indication that Nietzsche left us is the following: to know what thought is, it is necessary to take the body as a guide and not the soul or the intellect. And regarding cognitive machines: to reproduce the power of thought we must, rather than follow a logical or linguistic model, start with the relationship that perception and cognition maintain with passive syntheses. It is only within the body that we can find the multiplicity that constitutes forces that produce thought and subjectivity.

5.2.7

Thought and knowledge therefore essentially consist in a falsifying reduction of the innumerable diversity of facts to identity, analogy, and denumerable quantities. It is only under these conditions that judgment is possible. Judgment does not create the appearance of an identical case but works on the assumption that identical cases exist. "Thus there is in every judgment the avowal of having encountered an 'identical case': it therefore presupposes comparison with the aid of memory" (Nietzsche, *La volonté de puissance*, 60).

Memory, in turn, presupposes a more primitive function that assimilates and brings together different cases, making both itself and judgment possible. Judgment is only possible after a certain assimilation occurs from within sensations; similarly, memory is only possible if we constantly emphasize what is already familiar, as in déjà vu. "Before judgment occurs, the process of assimilation must already have taken place; thus here, too, there is an intellectual activity that does not enter consciousness, as pain does as a consequence of a wound. Probably an inner event corresponds to each organic function; hence assimilation, rejection, growth, etc." (60).

It is within the human body that we find the multiplicity of these functions, and therefore thoughts, because it is only there that the most ancient and most contemporary of organic becomings are incarnated and relived. The Nietzschean body is a coordination and hierarchization of an infinity of beings, an unheard of community of living beings. And there are as many consciousnesses as there are beings. What we call "the body" is actually a composition, a hierarchical coordination, of living beings, which should not be conceived as spiritual atoms but as beings that grow, struggle, strengthen, or decay. Within the body we find the most ancient organisms "continually growing or periodically increasing and decreasing according to the favorability or unfavorability of circumstances" (300). But we also find impressions and emotions of organic beings whose development occurs prior to the sense of the unity of consciousness. And then there is memory and imagination, which function independently of conscious control. This prodigious synthesis is called the body.

The consciousness of the ego is the last layer that is added to an already perfectly functioning organism: "The degeneration of life is conditioned essentially by the extraordinary proneness to error of consciousness. . . . But consider how heart, soul, virtue, spirit practically conspire together to subvert this systematic task—as if *they* were the end in view!" (300). Consciousness denies the multiplicity as well as the coordination and hierarchization that composes the human body, because it wants to appear as the sole origin of thought and ego.

For Nietzsche, we are a multiplicity that has been fabricated into an imaginary unity. And it is the intellect with its illusions—language, as we know, is its most advanced fetish—that has eliminated this multiplicity. Yet it is this multiplicity of beings, with all their contradictions, struggles, and wills to power, that constitute thought. The intellect operates only with a selection of simplified and easily accessible experiences. Even if there is a hierarchy and coordination between lower- and higher-order beings, the exclusivity that the modern tradition places on the role of consciousness risks leading us toward a false interpretation of ourselves. "The most important thing is: that we understand that the ruler and his subjects are of the same kind, all feeling, willing, thinking—and that, wherever we see or divine movement in a body, we learn to conclude that there is a subjective, invisible life appertaining to it. Movement is symbolism for the eye; it indicates that something has been felt, willed, thought" (301). The perspective offered by consciousness is not simply an error but is first of all a specific will to power that corresponds to the arbitrary takeover of certain cognitive functions. Yet consciousness and the intellect are very poor and attenuated compared to the multitude of consciousnesses and wills at work within us.

The human body is not only the synthesis of a potentially infinite number of organic beings—dependent and subordinated but, in another sense, dominant and endowed with voluntary activity—and their relationships of coordination and struggle. The body is also a composition of social instincts, because we have integrated ourselves within society and our naive and animal ego has been severely compromised by social practice. The sequence of the multiplicity of organic and social beings provides us with the composition, but also the condition, for the metamorphosis of the body. There is no opposition or contradiction between the individual and the collective, between the individual and the social, because the body is a multiplicity plugged into social multiplicity. "We carry within us a 'society' reduced to our own measure. . . . Olive groves and thunderstorms, even the stock exchange and newspapers, have become parts of ourselves" (285).

Social practices are working on us, not from the outside, but from within. There is no contradiction between the individual and the collective, because everyone is already a collectivity. There is a transition, reversibility, and struggle between these two forms of multiplicity. We must therefore reverse the priorities: consciousness must be humbled, since it is merely an instrument at the service of the body in its dual form as individual body and social body. The principle activity, according to Nietzsche, is the unconscious. "We must thus reverse the hierarchy: all consciousness is of *secondary importance:* what is *closer and more intimate* to us is not a reason, at least not a moral reason, to assess it differently. To take the *closest* for the *most important*, that is precisely the *old prejudice. We must change our way of thinking as to the principle of evaluation.* The spiritual must be taken for a semiotics of the *body*" (302).

5.2.8

I would now like to return to the comparison, outlined in the introduction, between Nietzsche's analysis of the body as a multiplicity of living beings, intellects, and social instincts and Karl Marx's analysis of the social body that represents the concept of the collective worker. In Nietzsche, we have a strong sense of the capitalist "machine" at work. This is a machine that deterritorializes bodies, subjects, and forms of production and representation, that liberates the forces and faculties within the human, but also liberates the forces and wills of nature, a machine that rearranges and reallocates them on another plane of immanence. Subject and object, consciousness and matter, are overwhelmed on all sides by flows that surge beneath and beyond the overcode: preindividual and prepersonal, supraindividual and suprapersonal forces. Postmodern capitalism restores to actuality a Nietzschean body, a reversibility between individual bodies and social practices. The collective within us and the collective outside us are becoming increasingly connected. Technological assemblages traverse and constitute these collectives as well as the preindividual and supraindividual conditions for the production of the real and of subjectivity. The development of these technological assemblages should be utilized to increase not intellectual activities but those of the body.

6

THE ECONOMY OF AFFECTIVE FORCES

6.1 THE ECONOMY AND THE PRODUCTION OF SUBJECTIVITY

6.1.1

Karl Marx points out the relationship between time and subjectivity as the key that opens us up to the enigmas of labor and the commodity as the "crystallization of time." Cinema, video, and digital technologies offer us another crystallization of time—new types of machines that, in contrast to mechanical and thermodynamic machines, crystallize the durations of perception, sensation, and thought. Before producing images and sounds, machines that crystallize time produce durations, and the mere possibility of reproducing them produces images and sounds. Machines that crystallize time are the motors of synthesis, contraction, and the creation of affective force. A "new kind of energy," a "powerful non-organic energy," is the matter upon which they work. The activity involved is directly related to a sui generis force, which is called the "feeling of effort" or attention.[1] Machines that crystallize time intervene directly in the processes of the production of subjectivity, since they deal with affects, perception, memory, language, and thought. I have already discussed all of these concepts. It is now a question of analyzing how affective forces and the production of subjectivities are at the heart of the process of the valorization of capital today. I will do this by linking Gilles Deleuze and Félix Guattari's analysis of capitalism with an analysis of the information economy.

We must abandon or completely reformulate a series of categories that have served to constitute a critique of political economy. I am thinking in particular of the concept of work (*travail*), which must confront, without nostalgia, the activities and affective forces that are solicited, produced, and consumed by new technological assemblages. Living labor—the critical linchpin of the concept of work—has been interpreted in Marxism as the expression of workers' subjectivity, which has made possible a profound renewal of the critique of political economy. But this concept is always directly linked to the qualification of subjectivity as worker subjectivity. The Marxist evocation of a global subjectivity as a defining characteristic of capitalism remains confined within this framework.

The concept of the production of subjectivity that has been put forth, with significant differences, by poststructuralist French philosophy allows for the introduction of a radical break within the Marxist definition of living labor while also recovering, from another point of view, Marx's original intuition: the subjectivity that has been "put to work" is indeed a certain kind of whatever-subjectivity (*subjectivité quelconque*), but only in the sense that it can no longer be described exclusively as worker subjectivity. In postmodern capitalism the Benjaminian distinction between work and perception, or the distinction made in the 1970s between work and affect, has been surpassed by a definition of generalized activity that, from the point of view of production, "may furnish surplus-value without doing any work (children, the retired, the unemployed, television viewers, etc.)."[2]

This passage poses formidable problems, since it is not a question of integrating work and subjectivity, work and language, work and affect, but of completely redefining the assemblages—the conditions of valorization and the production of subjectivity—in a world in which there are no longer any distinctions between "man-nature, industry-nature, [and] society-nature" or any "relatively autonomous spheres that are called production, distribution, consumption."[3] This redefinition demands a reconstructing of the concepts of free action and affective power, which have been perverted and mystified by both capitalism and the labor movement. Indeed, neither of these perspectives has been liberated from the theoretical and political subordination of the concept of power to the concept of work.

Even if I am forced to abandon most categories of the critique of political economy, I will not abandon the Marxist methodology; namely, the need to discover, beneath the categories of political economy, the genetic, creative, differential element that Marx defines as living labor. Here I argue that the concepts of language, communication, and information conceal and mystify the assemblages of the production of subjectivity as well as the affective forces that constitute them. Indeed, language and communication tend to enclose subjectivities, virtualities, and affective forces within the "faculties of the soul," but this time defined—and herein lies the ontological and political novelty—as intersubjectivity, as relation to the other.

Jürgen Habermas, Claude Shannon, and most linguists should be considered exactly as Marx considers the classical economists. Language, communication, and information are forms of spatialization of the activity of affective forces within the new conditions of capitalist accumulation. With language, communication, and information it is impossible to determine the genetic and malleable element, which alone can explain their constitution and evolution. "But where, for that matter, do they get the idea from that the socius can thus be reduced to the facts of language, and these latter in turn to binarizable, digitizable signifying chains?"[4] Deleuze and Guattari's *A Thousand Plateaus* not only recognizes that the production of value is no longer based upon the human component of labor but also provides the elements necessary for an articulation of the relationship between the production of subjectivity and the machines that crystallize time, beyond the categories of use value and exchange value.

6.1.2

Wage labor is productive (of economic value) only to the extent that it manages to capture and discipline desire and affective forces. Capitalism has always organized this capture through divisions: between factory and society, between worker subjectivity and whatever-subjectivity, between productive labor and unproductive labor, between labor time and the time of life, and between manual labor and intellectual labor. It is only under these conditions that the productive relationship between subjectivity, body, and time that capitalism gives rise to can be represented as the power of capital and wage labor.

The great transformation heralded by the struggles of 1968 lies in the fact that the relationship between desire, affective forces, and time no longer needs to pass through wage labor to produce wealth. The information economy shows us how capitalism itself, in its most advanced form, organizes the relationships between affects, desires, and time without passing through the discipline of the factory, but by capturing in open space the affects and desires of everyone—without distinguishing between productive and unproductive, between worker subjectivity and whatever-subjectivity—in order to give them the goal of the production of profit.

The analysis of capitalism developed by Guattari in the 1970s allows us to understand the increasingly important role that affective forces and machines that crystallize time play in the organization of the economy. He places at the center of his analysis two complementary aspects that have since been confirmed. First, contemporary capitalism is no longer limited to exploiting labor but society as a whole. "The notion of the capitalist enterprise has become inseparable from the whole social fabric, which is directly produced and reproduced under the control of capital. The concept of the capitalist enterprise should be extended to include the state apparatus, collective facilities, the media, the workplace, as well as many non-paid activities. In a way, the housewife occupies a workplace in the school, the consumer in the supermarket, and the viewer in front of the screen."[5]

Capitalism is no longer limited to exploiting labor time but also the time of life. To use two Foucauldian concepts, contemporary capitalism is defined as a biopower and a biopolitics. Putting life at the center of valorization implies—and this is the second aspect of Guattari's analysis—putting at the center of the production of value the affective forces that constitute it. "Capitalism claims to take upon itself the burdens of desire carried by the human race" (Guattari, *La révolution moléculaire*, 92). The organization of valorization thus invests not only economically recognizable values but also mental and affective values, the faculties of the soul, and the impersonal affects that are the foundation of the assemblages of production of subjectivity. "It is upon the basic ground of perceptual, sensory, affective, cognitive, and linguistic behaviors that the capitalist machinery is grafted, because individuals are equipped with modes of perception and the normalization of desire as if they were factories, schools, and territories" (91).

What capitalism manufactures is not only the flows of raw material, the flows of energy, and the flows of human labor but also the flows of knowledge and signs that reproduce affects, sensations, attitudes, and collective behaviors. The apparatuses for the production of subjectivity thus tend to identify themselves with the processes of wealth production.

According to Guattari, capitalism is defined by a process of deterritorialization of the real that can only be controlled and captured through asignifying semiotic machines and the flows of signs that are, in turn, deterritorialized. On the economic level capitalist power does not create discourse but only seeks to master these asignifying semiotic machines. Guattari refers primarily to money, to the organization of its circulation and to the asignifying grids of the stock exchange. The deterritorialized flows of subjectivity are therefore controlled and captured by the deterritorialized flows of money.

Contemporary capitalism is defined by a continuous enrichment of semiotic components and asignifying machines of capture, which are no longer limited to money and its derivatives. The putting to work of the socius and affective forces requires a specific machinery. It is in this sense that Guattari understands the development of what I am calling machines that crystallize time—television, cinema, and electronic networks—which perform the work of semiotization in a way that directly affects subjectivity. "Beyond fiat currency, credit, stocks, property titles, etc., capital manifests itself today by semiotic operations and the manipulation of power of all kinds, which involve information technology and the media" (97). The valorization of life requires machines capable of capturing affective forces as well as the nonorganic energy that constitutes life. With machines that crystallize time, the site of machinic integration is no longer limited to sites of production, but also extends to all social and institutional spaces: media, networks, collective facilities, and so on. My entire project here is to demonstrate how, through these new machines, capitalism can take control—beyond wage labor and monetized goods—of a multitude of quanta of power that were formerly encased within the local, domestic, and libidinal economies.

The novelty of Guattari's analysis lies in the fact that collective facilities—including the media and information networks—are not, contrary to what Louis Althusser thought, ideological state apparatuses. For Guattari, it is not a question of apparatuses for the reproduction of ideology but for the reproduction of the means and relations of production. The development of the last twenty years completely confirms this analysis and allows us to assert that the information economy has today turned these facilities into apparatuses for the direct creation of economic wealth. With the information economy, these facilities have even become the most dynamic and, quantitatively, the most important aspect of post-Fordist accumulation.

On the basis of Guattari's work, we can propose another important qualification of machines that crystallize time, which is closely related to the critique of their function in ideological reproduction. Machines that crystallize time operate not only through representations but also, and above all, through affects. To use a Bergsonian concept, these machines, through the work of production, conservation, and accumulation of duration, are grafted onto both "affective and representative sensations."[6] These machines function on a dual register, with asignifying semiotics (durations, intensities, affects) on one side and signifying semiotics (representations, ideas, feelings) on the other. This distinction is of crucial importance for not reducing machines that crystallize time to apparatuses of ideological reproduction as well as for understanding how they participate in the accumulation of the new kind of energy Bergson spoke of. It is through these two levels that machines that crystallize time directly engage with desire and affective forces in the information economy.

While signifying semiotics and machines engage with the global person and easily manipulated subjective representations (ideas and feelings), asignifying semiotics and machines assemble the infrapersonal and infrasocial elements of a "molecular economy of desire that is much more difficult to contain within stratified social relationships. By successfully putting these percepts, affects, and unconscious behaviors to work directly, capitalism takes possession of a work force, which far surpasses that of the working class in the sociological sense."[7]

A final remark about Guattari's reflection on capitalism brings us back to the fundamental functions of contraction and condensation in Bergson's work on memory consciousness and affective force, or pathos. In the deterritorialized world where things, to speak like Bergson, lose their solidity and stability and appear instead in the form of tendencies and flows, capitalism is described as the "integral of power formations" and collective facilities as "semiotic condensers" or "catalysts."[8] These terms—integral, condenser, catalyst—indicate how social and technological machines operate within the functions of memory and consciousness: preserving duration and intervening in the process of the bifurcation of time on the basis of the virtual-actual relationship in order to capture affective forces.

6.1.3

In *A Thousand Plateaus* there is a fundamental distinction made about the relationship between man and machine, depending on whether subjectivity refers to a "motorized machine" or a "cybernetic machine."[9] The book also distinguishes between two other concepts: "machinic enslavement" and "social subjection," which constitute two different forms of subordination (Deleuze and Guattari, *A Thousand Plateaus*, 451). "There is enslavement when human beings themselves are constituent pieces of a machine that they compose among themselves and with other things (animals, tools), under the control and direction of a higher unity. But there is subjection when the higher unity constitutes the human being as a subject linked to a now exterior object, which can be an animal, a tool, or even a machine" (456–57).

Capitalization pushes subjection to its most radical expression: the human is no longer a component in the machine but its user. We are not enslaved but subjected to the motorized machines of capitalism. The system of wage labor does not treat the human as a machine but trains it as a subject. *The worker is both subject and user of the machine.* And only on this basis can we understand the concept of alienation or Walter Benjamin's concept of the loss of experience. But the deterritorialization of contemporary capitalism restores and reinvents, through other technological apparatuses, a system of machinic enslavement that *A Thousand Plateaus* has traced back to imperial formations. "Cybernetic and informational machines form a third age that reconstructs a generalized regime of subjection: recurrent and reversible 'humans-machines systems' replace the old nonrecurrent and nonreversible relations of subjection between the two elements; the relation between human and machine is based on internal, mutual communication, and no longer on usage or action" (458). And *A Thousand Plateaus* highlights how this new configuration gives rise both to novel power apparatuses and to changes in the forms of accumulation and labor. The two elements are closely linked. This new enslavement involves apparatuses of standardization, modulation, and information that deal with language, perception, desire, and affects, which pass through microassemblages.

For its part, the form of accumulation is characterized by "a progressive increase in the proportion of constant capital" that determines a new kind of enslavement, which invests society as a whole and is directly linked to automation (458). The factory ceases to be the site of the determination of the conditions of production, which now intersect with the social conditions of the production and reproduction of society. Productive subjectivity in contemporary capitalism is no longer qualified as a subordinate subjectivity; it is, on the contrary, an arbitrary or whatever-subjectivity. This does not mean that subjectivity and society are not exploited. Whatever-subjectivity and society are the new conditions of capitalist accumulation.

> Surplus labor, capitalist organization in its entirety, operates less and less by the striation of space-time corresponding to the physicosocial concept of work. Rather, it is as though human alienation through surplus labor were replaced by a generalized "machinic enslavement," such that one may furnish surplus-value without doing any work (children, the retired, the unemployed, television viewers, etc.). Not only does the user as such tend to become an employee, but capitalism operates less on a quantity of labor than by a complex qualitative process bringing into play modes of transportation, urban models, the media, the entertainment industries, ways of perceiving and feeling—every semiotic system. (492)

In post-Fordist capitalism the processes of subjection are subordinated to the machinic enslavement with which they combine, but in entirely new ways relative to classical capitalism. Subjection, as we have seen, concerns global persons, representations, ideas, and feelings, while enslavement concerns more molecular and impersonal elements: affects, durations, infrasocial and infrapersonal elements. Enslavement knows neither subjects, roles, nor delimited objects. "It is precisely this that gives it a kind of omnipotence since it passes through the systems of signification within which individual subjects are recognized and alienated."[10]

Enslavement is called "machinic" because it is not centered primarily on human subjectivity, but rather on its conditions of production: flows of signs, social flows, and the most diverse material flows. But it can, as one possibility, recover subjection, which in turn changes its nature, since it is grounded upon the capacity of machines that crystallize time to reproduce the durations of perception, memory, and sensation. It is in this way that capitalism takes hold of beings from within. Their alienation, by way of images and ideas, is only one aspect of a general system of enslavement of their modes of semiotization, perception, and sensation, both individual and collective. Thus, one can be both subjected and enslaved to machines that crystallize time.

> For example, we are subject to television as long as we use and consume ("it is you, dear viewers, who create television . . .") The technical machine is the medium between two subjects. But we are enslaved by television as human machines as long as we are no longer simply users and consumers, nor even subjects who are supposed to "manufacture" it, but when we become its intrinsic components, its inputs, outputs, and feedback, which belong to the machine and no longer to the manner of producing or using it. In mechanistic enslavement, there are only transformations or exchanges of information, some of which are mechanical and others human.[11]

Viewers are simply "synapses," intervals, relays that convey information and ensure its circulation.[12] They are intersected by asignifying and signifying flows that exceed them on all sides. But simultaneously, through these same flows they are solicited to enter into processes of subjectivation that reintroduce representation as their unsurpassable form. In the postmodern communication system, the alternatives can be summarized as follows: (1) deterritorialization and enslavement in human-machine systems, in which viewers are literally intersected by flows of information and reduced to one of its machinic levels ("synapses"); and (2) reterritorialization as a subject that refers to an object, as consciousness ("dear viewers"). The media functions on the basis of the subjection-enslavement couple by continuously relaying us from one pole to the other.

If one can be simultaneously subjected and enslaved to these new technological assemblages, machines that crystallize time are also powerful vectors of subjectivation. Guattari's modelization, which contains these two alternatives, is based upon an actualized "plane of machinic interfaces" and a multiplicity of "virtual existential universes" whose functioning is paradigmatically expressed in the human-television relationship. Television, as a machine that crystallizes time, provides an insight into how these technological assemblages function at the heart of the processes of subjectivation. The human-machine relationship determines a specific plane of machinic interfaces open to crystallizations, which escape the complexity of actualized assemblages. Machines that crystallize time operate within the bifurcation of time, within the process of actualizing the virtual. "When I watch television, I exist at the intersection: 1) of a perceptual fascination provoked by the screen's luminous animation which borders on the hypnotic, 2) of a captive relation with the narrative content of the program, associated with a lateral awareness of surrounding events (water boiling on the stove, a child's cry, the telephone . . .), 3) of a world of fantasms occupying my daydreams."[13]

The heterogeneity of machinic assemblages is coupled with the diversity of the components of subjectivation that traverse the spectator. How can we preserve a relative sense of uniqueness? Guattari asks. The answer is, through a "ritornello" capable of freeing up processes of subjectivation within machinic, discursive, and actualized assemblages. "I would say that the ritornello is not based on elements of form, material, or ordinary signification, but on the detachment of an existential 'motif' (or leitmotiv) which installs itself like an 'attractor' within a sensible and significational chaos."[14] Subjectivity is an event-driven production in which television functions as a catalyst of the actualities and virtualities contained within the complexity of assemblages. Machines that crystallize time therefore operate at a much deeper level than ideology and its apparatuses. "It's not a matter of transmitting messages, investing images as aids to identification, patterns of behavior as props for modelization procedures, but of catalyzing existential operators capable of acquiring consistence and persistence."[15]

The interfaces of the human-television machine trace a plane of consistency in which the separations between form and content, sensation and intellect, and subject and object find a new continuity, a malleability that is simultaneously a power of metamorphosis and a creation or enslavement.[16] Around this plane traced by the recurrence and reversibility between human and machine, something fundamental plays out for both subjectivity and power. There is no need to patrol or discipline the collective assemblages of subjectivation according to the rules of productive labor. On the contrary, it is a matter of pursuing *the infinite modulation of affects and affections by seeking their differentiation.*

Bernard Cache analyzes our current transition toward an information economy as a transition toward a "nonstandardized mode of production." According to him, our everyday actions and gestures have irretrievably shifted from the realm of "tradition" or "law" to that of the norm. This process, which has long accompanied capitalist development, has now reached maturity. "The purpose of the norm is not to stabilize our movements; on the contrary, it is to amplify the fluctuations or aberrations in our behavior."[17]

Fashion is the paradigm for the new mode of production. "Fashion is the model of the norm. The rigid elements of our behavior are articulated with one another in order to produce increasingly variable configurations. As the turnover rate of capital accelerates, the modern gesture dissolves into variations" (Cache, "Terre meuble," 116). The dissolution of the rigidity of behavior goes hand in hand with the dissolution of the solidity of the object. "Modern objects are undermined by time.... Objects, which are those solid parts of our actions, are but a moment of densification in the folds of our behavior that is itself fluctuating" (116). Here we find again our double movement: a capitalist deterritorialization that dissolves the real according to speeds and rhythms, and a reterritorialization that functions according to densification-variation.

The Bergsonian ontology based on the continuity of the objective and the subjective—the variation and malleability of machinic interfaces that trace the plane of consistency of the production of subjectivity—corresponds to the ontology of the new economy. The recovered continuity of the universe is transformed into the continuity of the processes of production and reproduction: a new industrial smooth space. "Manufacturing techniques show the same variability as our behaviors. Precisely, digital machines and productive technologies in general allow for the production of an industrial continuum. From the mold we move toward modulation" (116).

6.2 THE INFORMATION ECONOMY

6.2.1

Now I would like to show how the definition of the production of subjectivity can entail one of the most relevant critiques of the new information economy. In light of the description of the economy's new norms, Guattari's modelization takes its full measure. Schizoanalytic modelization and the analysis of the information economy evolve naturally on different levels: the first defines an ontology of contemporary capitalism, and the second expresses the ideology of political economy. Mediations between the different levels are not yet sufficiently developed, but to grasp the activity beneath the information economy, the process of the production of subjectivity is perhaps more pertinent than the traditional concepts used in the critique of political economy.

The socioeconomic description of the information economy can be useful from another point of view: to provide a more determined framework for the capitalist machine that, here and now, is assembling the technological, social, and semiotic flows of the "post-media era." There is an urgent need to report on the development of the machines that crystallize time, with regard to not only anthropological and aesthetic changes but also, and above all, the capitalist initiative, which in this area has already made considerable progress.

I would like to begin with a quotation from Robert Fogel, the 1993 winner of the Nobel Prize for Economics: "The technical-industrial system is no longer organized around matter or energy, but around the structuring of time."[18] After reading Bergson and analyzing machines that crystallize time, we find that this statement from an economist takes on a very special meaning. For Fogel, "the object of information and communication technologies is, above all, to contract time, that is to say, to decompose it ever more finely."[19] Time has lost the function of measurement that had been attributed to it by political economy, and even if it remains spatialized, it opens up within the economy a terrain of analysis in line with the direction being proposed here.

Affective forces, machines that crystallize time, the new relationship between humans and machines, human and nonhuman semiotic assemblages, subjectivities and protosubjectivities, and the virtualities that double these different moments are put to work by the organization of the information economy. The reversibility of humans and nature, the artificial and the natural, is the new condition for the dynamics and development of the market.

Let us begin by describing what American observers have called "techonomics," or the attention economy. Its characteristics can be summarized as follows: (1) The motor of techonomics is the relationship with the customer. We thus move from an economy of supply to one of demand, from a capitalism of production to one of products. Consumption is no longer relegated to the realization of the value of the commodity. On the contrary, the relationship with the customer becomes the center of production, the point around which the whole cycle of valuation is structured. This does not entail the disappearance of industrial labor, but rather a deep restructuring and, above all, a radical subordination to the relationship with the customer. (2) The technological condition for this new form of production is represented by digital technologies. The latter transform telephone networks into telematic networks and transform terminals from simple, passive reception devices (the television) into truly "smart" devices (the computer) for processing information: contracting and dilating, delaying and accelerating. These technologies transform networks that were mere carriers of information into real markets for the production and exchange of "services with a very high added value." The coupling of television with the digital is the condition for the penetration of new apparatuses into society. (3) These first two conditions of techonomics determine the fact that the production of value takes place in the creation of services—"the added value of products and services is on the side of software and information content"—and in the social conditions (training, welfare, collective facilities, and so on) that make it possible to relate with this new productive space.

The assertion that only customer service is a source of value must be interpreted as a new relationship that the information economy maintains between affective forces and the socioeconomic conditions of their existence. The information economy functions as an "apparatus of capture," which appropriates the production of relationships, affects, and subjectivities that society creates independently of it.

Indeed, the information economy considers industrial labor and all the conditions that make possible the reproduction of society and the population (including knowledge, desires, languages, and affects) to be disposable natural resources, similar to how raw materials and populations were regarded in the eighteenth and nineteenth centuries. The information economy does not produce labor and the conditions for the reproduction of society, but they constitute many presuppositions of its operation.

Society as a whole produces, creates, and innovates, but it is only here that the realization of surplus value becomes visible. Computerization is the other face of the process of the financialization of the economy and society. However, what the information economy regards as "natural data" must be ensured and driven by specific economic and social policies. The education of the population, the capacity to constantly adapt to market fluctuations, the capacity for innovation, the right conditions for reproduction, the forms of semiotization, and the forms of the redistribution of income all become presuppositions for the operation of the information economy. These economic characteristics had been isolated in the 1980s, but according to American observers, a true rupture in the organization of the American economy only occurred at the beginning of the 1990s.

In the 1980s US economic growth was energized by the boom in services—health services, industry services, personal services, legal services, and so on—so much so that, at the time, there was talk of a transition from an industrial to a service economy. The disadvantages of such a transformation were many: service jobs were low wage compared to industrial jobs, lower in productivity, and difficult to export. This helped to multiply the declarations of a US economic decline. But in reality, these years correspond to a gestation period of even deeper changes that only manifested in the 1990s.[20] "More quickly than most people can realize, it is possible to verify that the growth of the economy is not driven by 'services,' but by the computer, software, and telecommunications industries. According to the Department of Commerce, spending (industrial and consumer) on high-tech equipment accounted for almost 38% of growth since 1990."[21]

But let us return to the characteristics of this new economy. More specifically, techonomics is defined by digital technology whose speed of development has immediate consequences for the profitability of investments. Indeed, there is an inverse relationship between the power and performance of these new technologies and their selling prices. This is what the Americans call "the technology paradox." The power of electronic chips doubles on average every eighteen months, while prices continue to fall. The major computer and electronic manufacturers have experienced a deterioration in their profitability and investment capacity because of the technological performance of the digital components of their products. And this fact takes on particular importance because, in the last five years, chip technology has crossed an invisible threshold, assuming a central role in the world economy.

According to the Americans, the evolution of digital technology shows that the production of added value in hardware tends toward zero and that, ultimately, this technology will be free. What the French state did with Minitel, by providing free support, industries are beginning to practice with their product lines. But software developers may find themselves in the same situation. Just like computer manufacturers, the software industry is exposed globally to severe price competition. Lower support prices resulting from the drop in component prices have forced software vendors to lower their prices, often dramatically. Only a monopoly position like Microsoft's can mitigate this trend. "Sooner or later, this fall in prices will reduce the value of almost every component of hardware and software. Then value will only be generated through a long-term relationship with the customer, even if this means giving away the first generation of products."[22]

Also according to the Americans, industry has entered a new phase: mass customization. This new phase will determine a new mode of production that will result from a synthesis of the productive and organizational innovations of the last two decades—"total quality management and tight flow organization"—but controlled and driven by the relationship with the customer. The customized products of new companies will be produced as quickly and cheaply as the standardized mass products of the Fordist mode of production, of which the Japanese have exploited all the reserves of productivity. The difference with tight flow organization is that "companies will not sell products, but they will sell bonuses to their customers."[23]

The economic calculation is itself completely disrupted. The price of the product will not be determined by adding up the costs of its parts, but by the "know-how and services that a company manages to accumulate in order to ensure the final satisfaction of the customer."[24] American observers insist that it is no longer enough to put the most efficient products on the market. Now the problem is the creation and management of customer relations. It is precisely this relationship that is the source of differentials in productivity. Therefore, the goal of this new mode of production is to connect customers, companies, and suppliers to a sort of hyperefficient confederation. Signs of crisis in the old way of production were already visible in the mid-1980s, and this is almost paradigmatic for the electronics industry. Indeed, this industry seemed to have come to a dead end due to the already-noted relationship between the power of technological performance and the price. Many experts at the time thought the only way to stay in the market was to mimic the vertical integration of major Japanese conglomerates in order to produce components, launch new products, and win the price war by lowering costs. But by the early 1990s another strategy was beginning to emerge. Instead of focusing solely on the product, companies started to target architectures that integrate a multitude of components into an overall operation. This overall architecture is profitable, but its implementation requires a real-time connection with consumers, flexibility, and the ability to anticipate evolutions in both the market and technology. "This approach requires intimate knowledge of customer needs, competence in forecasting technological change over three to five years, and a willingness to look beyond easy opportunities offered by current market conditions."[25]

This high-tech economy completely shifts business priorities and entrepreneurial know-how. Techonomics is also referred to as the attention economy. If the core of valuation is the long-term relationship with customers, if technological changes and consumer desires are not standardized but rather in constant evolution, how can customers be retained, and how can we attract their attention? "If attention is the most valuable resource for the 'free-tech economy,' then the interface between human and machine already consumes almost three-quarters of the development work of electronic products."[26] According to the president of Syntec, 80 percent of new operating systems are nested within client-server architecture. Here, it is not computing as such that is at the heart of capitalist strategies, but rather this client-server architecture: "We intervene when there is integration to be done in restructuring systems. And the integration of systems revolves around client-server architectures, providing microphones, networks, and software."

The final cost of the components is marginal relative to the price of the final package, which includes services as well as the possibility of future changes in the customer relationship. It is here, as I have already mentioned, that the Japanese have gone astray, even though Nintendo was one of the first to understand these developments. "There's no future in looking for value in hardware," according to the president of Nintendo.

Edward McCraken of Silicon Graphics pushes the definition of techonomics even further. According to him, techonomics is not a business of commodities. The information economy would fall into a classical economic logic if price were the key to competitiveness in the mass market and if efforts were made to be the leader through low costs rather than through innovation. Under these conditions, the economy would indeed fall into the business of commodities. However, the theory of innovation reaches its maturity in these enterprises and profoundly modifies the conditions for its realization, since, on the one hand, innovation is completely determined by the relationship with the customer, and on the other hand, the turnover rate of these technologies is much higher than that of traditional industry. "There has never been a commodity market with a rate of technological change like the one that exists in the computer industry. . . . When performance doubles, the paradigm shifts because everything has changed: the software, the packaging of the product, the way customers use it. If we are not willing to try to shift the paradigm, then the products quickly become 'commodities,' as has happened with the personal computer market."[27] Companies in the business commodity model respond to market chaos by trying to anticipate or keep up, while the philosophy of companies that belong to this new configuration is that "the key to gaining a competitive advantage is not to react to chaos, but to produce it. And the key to being producers of chaos is to be a leader of innovation."[28]

6.2.2

How should we understand and, more importantly, critique this new political economy centered on the relationship with the customer? The concepts of enslavement and subjection can certainly be useful in this regard, because to what kind of subjectivity is the production of final satisfaction addressed? Certainly not to a subjectivity of linguistic theory or of the communicative act, constructed upon the anthropomorphism of the subject, the ego, and the alter ego. No doubt it is addressed to a machinic subjectivity. But how is this subjectivity addressed? Surely not through a logocentric enunciation, but rather through forms of "machinocentric" enunciations.[29] As we have seen, machines that crystallize time operate at the heart of sensibility, memory, intellectual labor, and language. Their flows and their semiotics are the conditions of any process of subjectivation. Our clientele is a machinic clientele constructed by the recurrences and reversibilities of the human-machine relationship, according to forms of subjectivation whose machinic enslavement and subjection designate our becomings.

What relationship does this mass customization take, from an economic point of view? It is certainly not a relationship between supply and demand constructed according to the connections and circulations of standardized merchandise, as in Fordism. Here mass customization is instead an event-driven production that functions as a converter of the actualities and virtualities contained within the complexity of productive, social, and technological assemblages. The production of this particular type of customer-subject is constituted by flows and semiotics that put affective forces to work. Subjectivity is really taken in this machine of techonomics. Language, communication, and their logocentric enunciations always keep us within this new world in the same way that the categories of political economy always keep us within the divisions between work and affect. The new political economy clarifies very well what I mean when I say that the concept of language proposed by the linguistic turn is absolutely insufficient to account for techonomics. "The network industry is moving from a domain in which the human need to communicate between individuals, mainly between individuals, to that of computer to computer communication provided by digital data transmissions. . . . Industry has moved from the domain of human space-time to that of the computer."[30] It is in this sense that Guattari prefers the concept of assemblage, for which human semiotics have no privilege, to a concept of language that is still based upon "man" and the imperialism of the signifier. The subordination of the whole of society to valorization highlights the complexity of its regimes of signs and relativizes that which the linguistic turn considers exclusively.

It is not only a matter of recognizing that all social machines—family, school, politics, work—operate at the heart of subjectivity. Rather, we must understand that this new temporal ontology and its technological assemblages completely redefine the conditions of existence and the production of social machines, since on the plane they draw, the distinctions between the social and the economic are no longer relevant. If as we have seen, video was the first technological assemblage to introduce this new plane of immanence, it is only with digital technology that this plane acquires its full power. Habermas's theory of communicative action, just like the linguistic turn, is incapable of accounting for how computers operate and the activities they involve: how can we distinguish between instrumental and communicative action when we work and communicate with the same machine? What becomes of speakers when dialogue is not only technologically polyphonic but also immediately machinic?

The object of the new economy is precisely the production of subjectivity, enslaved and subjected, under conditions determined by this machinic plane. And it aims at subjectivity in general, a whatever-subjectivity. It is indeed society in its complexity that is exploited, that is engaged in the relationship of valorization. Techonomics makes it clear that the fundamental question is one of the relationship with the customer, to which everything else is subordinated. The social assemblage in its totality is targeted.

6.2.3

Let us return to the information economy. The point I would like to emphasize now is that the production of services is not limited to client-servicer architectures but must also include the production of content. Artistic production, and more generally cultural production, has undergone one of the greatest transformations in its history. The situationists had foreseen this development thirty years ago, alongside many other issues in this field. The heralds of the information society and cyberspace have discovered nothing new. On the contrary, they have often overlooked certain details, in particular the fact that the evolution of this new society has always taken place within the abstract capitalist machine. "Culture, which has become an integral commodity, must also become the star of the society of the spectacle. Just like railroads in the second half of the previous century and the automobile in the first half of this century, culture must play a leading role in the development of the economy in the second half of this century."[31]

To this statement of the situationists, we could simply add that culture is the driving force of the economy in its assemblage with communication systems and machines that crystallize time. It is therefore necessary to extend the concept of software to the production of any cultural content, which will ultimately do the footwork of the information economy.[32] The chairman of Viacom, Sumner Redstone, claims in the *Financial Times* that "software is king, was the king, and will always be the king." And by software, he means the production of films, television programs, and books.

The labor market of cultural production is certainly one of the most dynamic. In France, according to the National Institute of Statistics and Economic Studies and the Ministry of Culture, the number of professionals has increased by 36 percent over the last ten years, ten times higher than that of all other social categories. This represents 1.7 percent of all jobs, which is equivalent to the automotive industry. There are an estimated 265,000 people in arts-related professions. If we add to this technical and administrative professions, cultural employment can be estimated at 377,000 people. This labor market has developed at or above the level of the production of computer and telecommunications equipment and has simultaneously entered into a violent process of segmentation and hierarchization. But the most decisive transformation is that of the information economy. Artistic activity has been completely integrated into capitalist valorization.

This phenomenon started gaining momentum at the beginning of the century and is now fully realized. The integration of artistic activity is of paramount importance for the information economy for two fundamental reasons: (1) In order to valorize hardware, the content becomes strategic. In this sense, the software must integrate all the data and factors that belong to the organization of behaviors. (2) The current development of information technology has reached unprecedented levels, while culture and society may not keep pace. The gap between culture and technology, between creativity and knowledge, must be quickly absorbed, and to this end the education of the sensible becomes strategic. It is therefore necessary to discover and utilize a new figure of invention and experimentation capable of acting in parallel on both axes, a figure that can be defined as an "artist-engineer" of the "new renaissance."[33] This is because techonomics, with new concepts and models, veils the fact that it traces a new smooth space on which it can exploit free action.

6.2.4

The complexity of Guattari's concept of modelization highlights the scope of the field of confrontation as well as the alternatives that are opened up in the production of subjectivity. The only thing that can be asserted with some certainty is that the flight of wage labor from the discipline of the factory has forced the system to confront not only the question of labor but also the more general one of the production of subjectivity. However, the recognition of subjectivity is henceforth the recognition of a whatever-subjectivity without any qualifications that connect it directly or indirectly to work. It is therefore possible to reinterpret the entire history of capitalism as an opposition between the concepts of labor and the production of subjectivity in order to reconnect, beyond all the mystifications of the workers' movement, to the concept of power.

A Thousand Plateaus proposes a theory that can be useful for understanding this opposition, which traverses the history of capitalism. In fact, Deleuze and Guattari produce two ideal models: "Work is a motor cause that meets resistances, operates upon the exterior, is consumed and spent in its effect, and must be renewed from one moment to the next. Free action is also a motor cause, but one that has no resistance to overcome, operates only upon the mobile body itself, is not consumed in its effect, and continues from one moment to the next."[34] *A Thousand Plateaus*, like Marx's work, highlights the political and social centrality of the work model for nineteenth-century capitalism, but unlike Marx, this theory always and immediately refers to whatever-subjectivity; that is, subjectivity without any "capitalist" qualification. The opposition is no longer between labor and living labor, but between the work model and free action. This completely shifts the problem and allows us to rethink the opposition that lies at the foundation of capitalism according to the alternative: work-power or work-activity. Capitalism works to impose "the work-model upon every activity, translate every act into possible or virtual work, discipline free action, or else (which amounts to the same thing) relegate it to 'leisure,' which exists only by reference to work" (Deleuze and Guattari, *A Thousand Plateaus*, 490). Free action is subjected to the work model through a double movement. On the one hand, labor appears only as a correlate of the constitution of surplus labor. That is, the theory of value can only be a theory of surplus value. On the other hand, "labor performs a generalized operation of striation of space-time, a subjection of free action, a nullification of smooth spaces, the origin and means of which is in the essential enterprise of the State, namely, its conquest of the war machine" (490–91).

I will make two remarks inspired by this text, both of which are hypotheses about labor. First, *A Thousand Plateaus* demonstrates how, already in Marx, surplus value ceases to be localizable in and by the exploitation of labor. "It gave him a sense that machines would themselves become productive of surplus value and that the circulation of capital would challenge the distinction between variable and constant capital. In these new conditions, it remains true that all labor involves surplus labor; but surplus labor no longer requires labor" (491–92). Second, *A Thousand Plateaus* puts forward a more interesting hypothesis for developing a critique of the information and communication economy: capitalism recreates the conditions for the exploitation of free action without passing through its subjection in work. As in primitive accumulation, capitalism must enslave free action, activity as such. Postmodern capitalism draws a plane of consistency beneath the labor-capital division and determines, on the contrary, a continuity between "molecular" and "cosmic" within which capitalist divisions and subjectivations follow other cartographies and other forces. "It is as though, at the outcome of the striation that capitalism was able to carry to an unequaled point of perfection, circulating capital necessarily recreated, reconstituted, a sort of smooth space in which the destiny of human beings is recast. . . . The multinationals fabricate a kind of deterritorialized smooth space in which points of occupation as well as poles of exchange become quite independent of the classical paths to striation" (492).

This new smooth space, which capitalism is obliged to reconstitute, is the one that emerged in the cybernetic human-machine assemblage. It is the space of recurrence and reversibility between natural and subjective conditions. This is the new temporal ontology. *A Thousand Plateaus* concludes its analysis by discussing the categories of "constant and variable capital, and even fixed and circulating capital" within the critique of political economy (492). According to this text, these distinctions no longer fit with the conditions of contemporary capitalism. "The essential thing is instead the distinction between *striated capital* and *smooth capital*, and the way in which the former gives rise to the latter through complexes that cut across territories and States, and even the different types of States" (492).

Under the threat of being overtaken by the new relations of power that are constituted in and against free action, it is necessary to accept with utmost rigor the displacement effected by the capitalist organization of society and be bold enough to invent new categories that correspond to it. "These phenomena confirm the difference between the new machinic enslavement and classical subjection. For subjection remained centered on labor and involved a bipolar organization, property-labor, bourgeoisie-proletariat. In enslavement and the central dominance of constant capital, on the other hand, labor seems to have splintered in two directions: intensive surplus labor that no longer even takes the route of labor, and extensive labor that has become precarious and floating" (469; translation modified).

In conclusion, I would like to highlight three fundamental displacements that the theory of the production of subjectivity addresses in relation to formal subsumption, displacements that open up a completely unknown terrain: from work to free action, from the subject to the production of subjectivity, and from technics to machinics. These three points reciprocally presuppose one another and summarize the theoretical displacement I have addressed specifically with the analysis of machines that crystallize time.

7
THE CONCEPT OF COLLECTIVE PERCEPTION

7.1

To conclude, I would like to revisit the themes that have been developed so far in relation to the work of Walter Benjamin and his concept of collective perception, which might revitalize, on a more political ground, the thesis already put forth. For Benjamin, though not for Henri Bergson, "the way in which human perception is organized—the medium in which it occurs—is conditioned not only by nature but by history."[1] The encounter between these two points of view will provide us with provocative ideas for reflecting upon current events. The value of Benjamin's methodology lies, I believe, in the fact that it directly connects the mechanization of labor and of perception, the collective forms of production and of perception, the shock produced by the assembly line and by montaged images, and the transformations of the commodity form and the introduction of technologies of reproduction of the work of art, as well as the crisis of the concept of art itself, the work, and the author. And all this is grounded upon cinema as an adequate medium for the socialization of the forms of perception that capitalism establishes.

Nonetheless, I will not adopt Benjamin's position regarding the concept of the technological reproduction of the work of art, which contains many ambiguities. Benjamin oscillates between the analysis of the automatic reproduction of the work of art, its standardized production, and the analysis of the specific temporalities of capitalism. It seems to me that the intuition that consists of linking the critical mutations of perception, of memory, of the processes of subjectivation, of automation and time, has not been explored with rigor down to its ultimate conclusions. For example, automation and time run parallel, without getting reciprocally involved, contrary to what I have tried to demonstrate.

Benjamin's analysis of collective perception is structured around the time-memory relation. The man of the city lives within the "spleen," unable to free himself from the fascination with the flow of empty time. The Baudelairean ideal, interpreted by Benjamin as an anticipation of the metropolitan type, responds to the loss of experience with an appeal to involuntary memory, a repository of images from a past life. The poetics of Charles Baudelaire, according to Benjamin, can be summarized in the attempt, always doomed to failure, to plunge the image within the recollection of involuntary memory.

The destruction of involuntary memory is the work of information that requires consciousness to respond with an intellectual shock, defined by Benjamin as the predominant form of sensibility in the industrial age. Consciousness is obliged to defend itself against such shocks and develops a form of voluntary memory that responds to stimulations by mechanical reflexes. It may be useful to mention Benjamin's reading of Bergson's *Matter and Memory* here to highlight the fundamental differences between his interpretation of Bergson and mine. Benjamin reads the work of Bergson within the opposition between the time of tradition (involuntary memory) and the value-time of capitalism (voluntary memory) that he himself has established. With his concept of memory, Bergson attempts to recover immediate experience and therefore opposes the habitual modes of experience specific to the industrial age. My interpretation relates Bergsonian memory not to the time of tradition but to the empty time of capitalism, time liberated from its subordination to the movements of the universe and the soul, and to its reversal as power-time, the time of creation. The concept of virtual memory can help us clarify the difficulties and ambiguities of the concept of Jetztzeit (the "messianic present" and "dialectical image"), which Benjamin at the end of his life defined as an alternative to both the empty and homogeneous time of information and the (impossible) restoration of the time of tradition.

Bergson, like Baudelaire, plunges the image into recollection but thereby discovers a more profound memory, an ontological memory that is the foundation of psychological and social memory. It is interesting to note that, for Benjamin, the conditions that provide access to the past, to social consciousness—not to a particular psychological or historical past but to the virtual past, to nonchronological time—are those that allow for time as such to reveal itself to individual consciousness: the scrolling of life before the eyes of one in danger of death (hanged or drowned, as in Bergson)[2] or the sudden emergence of a memory. "Historical materialism is committed to capture an image of the past as it comes about unexpectedly and at the very moment of supreme danger."[3]

The present—the most contracted form of the past—when released from its subordination to the need to be useful for purposive action, and from its subordination to the time of everyday banality, leads us to experience time itself. The rupture of the unification of time and the image within the sensory-motor schema is represented for Benjamin by the revolutionary act, which blocks the flow of value-time. The fundamental task of the revolution—the double articulation of destruction and construction—is announced in "On the Concept of History" as a task that directly concerns time. "Thinking involves not only the movement of thoughts, but their arrest as well. Where thinking suddenly comes to a stop in a constellation saturated with tensions, it gives that constellation a shock, by which it is crystallized" "[and] can be seized only as an image that flashes up at the moment of its recognizability." "The historical materialist approaches a historical object . . . [by recognizing] the sign of a messianic arrest of happening, or (to put it differently) a revolutionary chance in the fight for the oppressed past."[4] It is only under these conditions that we can break through the empty and homogeneous continuity of value-time and grasp the singularity of an era as well as that of a life. In this quote Benjamin emphasizes that the continuity of value-time is exploded by nonchronological time, by which he means a primary past that is true for all times. The messianic present is a time that contains all times (the entire past), because it is itself the most contracted form.

It seems that Benjamin often hesitates between one text and another, between a foundation of time in a "past that conserves" and another in a "present that creates." The double foundation of time we find in the Bergsonian concept of virtual memory does not seem sufficiently articulated in Benjamin. Even if the opposition between historical times is precisely defined, it is not so with regard to the ontological conditions of time. The present as event, as opening up nonchronological time, reoccurs continuously, and often alternately, between these two forms of virtual-ontological memory. And it is this recurrence that gives a particular tone to Benjamin's work; for example, between the "new barbarian" who, under the capitalist conditions of memory's absence, must seize the opportunity to break free from the deceptive opacity of his inner life and the new religious type who, like the Messiah, should liberate and redeem the past for all those exploited and defeated in history.

The difficulties and ambiguities of the concept of Jetztzeit are probably due to Benjamin's original attempt, absent in the work of Bergson, to articulate historical time in its ontological form. With Benjamin we find a thematization of historicosocial conditions that announce and prepare for the reversal of metered time and power-time, a thematization that we must infer from Bergson. The mutation introduced by the technological reproduction of the work of art, Benjamin argues, determines the conditions for an awakening of the political role of the image and time. But despite this deep problematization of time, the precise relationship that Benjamin establishes between mass reproduction and the reproduction of the masses may obscure the temporal processes of industrial production-reproduction, starting with cinema.

Cinema is not definable, first of all, as a serial process of reproduction of the unique existence of the work of art, but more fundamentally as an apparatus that introduces movement and time to images. It is an automatic apparatus for the crystallization of time, a motor that produces and reproduces the syntheses of time, as I have tried to explain. Benjamin essentially understands the technical reproduction of the image as being the reproduction of a copy whose model might be the printing press, whereas for us technology reproduces time. The concept of machines that crystallize time is intended to demonstrate how capitalism operates an automatic reproduction of time that is the raw material of perception, memory, and subjectivity. The concept of the mechanical reproduction of the work of art assimilates these technologies to mechanical technologies. I have attempted, in contrast, to highlight their originality as the technologies of time.

With this objection, we can now return to Benjamin's work and to three points: (1) The socialization of the forms of perception and reception, which finds in cinema its debut and in the masses its subject. The process of the production and individuation of subjectivity is organized by technological (machinic) apparatuses in the same way as in the process of material production. (2) The collective form of perception determines a radical transformation in the forms of production as well as in the reception of works of art. The transformation of the exhibition value of the work of art is not only due to the industrialization of the production of works but, above all, to the activity of the masses who want to reduce the distance that separates them from the work. The form in which this approach manifests is that of collective perception, constituted in distraction and entertainment. (3) Collective perception transforms the audience into experts. This transformation is directly linked, for Benjamin, to the forms of socialization and cooperation that are constituted in the process of work. The transformation of the public and collective worker are two sides of the same process. Indeed, Benjamin sees in the collective forms of cinematic production the purest form of exceeding the capitalist division between manual and intellectual labor. These themes are so contemporary that I will use them to draw, very briefly, the subsequent steps of the development (from television to digital networks) of collective perception and machines that crystallize time.

7.2

Let us revisit these three points in more detail and see how the changes in capitalism and class conflict affect collective perception, the concept of the public, and the very nature of labor. The alignment of reality with the masses is, according to Benjamin, a phenomenon of the utmost importance that affects all fields. The masses constitute the matrix from which new attitudes are generated with respect to perception, sensibility, and art. The technical reproduction of art alters how the masses react to it. Cinema, the first form of collective perception adequate to the masses in the era of big industry, verifies and defines these new attitudes. Their main characteristics consist in practices that tend to sever the distance that the work of art had typically established with respect to its consumers. In this form of renewed perception, emotional pleasure is intimately conflated with the attitude of the expert. The close relationship between a critical attitude and simple pleasure is a symptom, for Benjamin, of the social importance of an art form. The reception of cinema, with its intrinsically collective form, differs from the reception of painting in churches during the Renaissance. We might therefore locate its antecedent in ancient epic poetry. The reception of the masses is characterized not only by its collective manifestation but also by the fact that it develops into distraction and entertainment. This attitude is motivated by the masses' desire to "get closer to things"[5]—to appropriate, know, and experience them, removing any traces of their aura, which as we shall see is an aura of time and power.

I will read the loss of the aura, not as a unilateral capitalist process, but as a manifestation of the active intervention of social subjects. I will stay close to Benjamin's methodology, which is based on the dual nature of commodities, the driving force of this transformation. Reception as distraction and entertainment is radically opposed to perception as contemplation and recollection. "A person who concentrates before a work of art is absorbed by it; he enters into the work, just as, according to legend, a Chinese painter entered his completed painting while beholding it. By contrast, the distracted masses absorb the work of art into themselves. Their waves lap around it; they encompass it with their tide."[6] Contemplation establishes a distance between the work and its consumers that the masses refuse, because it bears a different temporality, a different sensibility, a different attitude toward the world.

7.3

I would now like to compare Benjamin's commentary with one of Mikhail Bakhtin's texts, in which he beautifully demonstrates that this attitude toward distance directly concerns time. Surprisingly, this takes us back to the problem of machines that crystallize time and to the syntheses that constitute them. Bakhtin reads the development of and struggle between literary genres as an attempt by minor, comic, and popular genres to reorient "towards the future," as the expression of a sensibility that feels closer to the future than the past. In attempting to overcome "contemporary reality," the high literary genres constitute, according to Bakhtin, a wretched present that passes and flows, a "life without beginning or end."[7] The main point of the artistic and interpretative evaluation of high genres resides, according to Bakhtin, in the absolute past, in memory, in a present that remains, in its flow, always unfulfilled and devoid of essence. "The interrelationship of times is important here: The valorized emphasis is not on the future and does not serve the future, no favors are being done it (such favors face an eternity outside time); what is served here is the future memory of a past, a broadening of the world—of the absolute past, an enriching of it with new images (at the expense of contemporaneity)—a world that is always opposed in principle to any merely transitory past). In the already completed high genres, tradition also retains its significance."[8] This hierarchy of time refers directly to that of power. The idealization of the past is a formality. All outer expressions of the dominant force and truth are organized in the categories of the past, distance, and memory, in "time as a closed circle," as Bakhtin says.

However, in popular comedy the present, the "me in person," "contemporaries," and "my time" are subject to ambivalent laughter, at once jovial and destructive. To the absolute past (the gods, demigods, and heroes) we oppose a present that is going toward the future, to distance we oppose "free familiar contact," and to the fullness of the past we oppose the violation of the present. It is here, according to Bakhtin, that new attitudes are born toward language, speech, representation, and more generally, power and tradition.

Benjamin's insights on perception as distraction and entertainment, on the desire of the masses to get closer to things, remind us of those secular carnivalesque attitudes about time that are, according to Bakhtin, the source of modern literature. We, on the other hand, find ourselves in the cinema, the origin of modern art.

> It is precisely laughter that destroys the epic, and in general destroys any hierarchical (distancing and valorized) distance. As a distanced image a subject cannot be comical; to be made comical, it must be brought close. Everything that makes us laugh is close at hand, all comical creativity works in a zone of maximal proximity. Laughter has the remarkable power of making an object come up close, of drawing it into a zone of crude contact where one can finger it familiarly on all sides, turn it upside down, inside out, peer at it from above and below, break open its external shell, look into its center, doubt it, take it apart, dismember

it, lay it bare and expose it, examine it freely and experiment with it. Laughter demolishes fear and piety before an object, before a world, making of it an object of familiar contact and thus clearing the ground for an absolutely free investigation of it. Laughter is a vital factor in laying down that prerequisite for fearlessness without which it would be impossible to approach the world realistically. As it draws an object to itself and makes it familiar, laughter delivers the object into the fearless hands of investigative experiment—both scientific and artistic.[9]

The comic genre is, for Bakhtin, the expression of an absolutely specific temporality. The function of memory is minimal: we laugh to forget. The time of popular comedy destroys the distance of the past and opens the time of indetermination, of infraction, with regard to creation. This temporality is based upon the future rather than the past. Its being requires a continuation in the future, and the more it advances toward this future, the more it becomes indeterminate, unfulfilled, open to the unforeseen creation of the new. It involves a present that is not only sociohistorical but equally abundant with the virtual. "It is precisely the zone of contact with an inconclusive present (and consequently with the future) that creates the necessity of this incongruity of a man with himself. There always remains in him unrealized potential and unrealized demands. The future exists, and this future ineluctably touches upon the individual, has its roots in him. An individual cannot be completely incarnated into the flesh of existing sociohistorical categories."[10] Bakhtin situates the artistic incarnation of this joyous overflow of time among the masks of the Italian commedia dell'arte. Harlequin and Pulcinella are the heroes of free improvisation, an always contemporary vital process, indestructible and forever renewing itself. Giorgio Agamben, in an article on cinema, made these characters the custodians of an artistic practice in which there is a "mixed power and performance that escapes the classifications of traditional ethics."[11]

These admirable Bakhtinian pages appear to me an example of the struggle over time. The two forms of temporality that underlie pure or virtual memory in Bergson (the past that preserves and the present that creates) become, in Bakhtin's hands, an affront to existential forms, processes of subjectivation, artistic practices, and the constitution of society and its purpose. And this refers back, as suggested by Agamben, to the concept of power (as power-time) and to two radically opposing ethics. Benjamin highlights this in his description of new attitudes that the masses adopt with regard to cinema and, more generally, with the form of reception of works of art. Theodor Adorno, in his correspondence with Benjamin, responded that the laughter of the cinema spectator is not revolutionary but, on the contrary, is imbued with the worst type of bourgeois sadism. Indeed, this laughter is not the full and autonomous carnivalesque laughter of Bakhtin, because it is caught within the dialectic of the dual nature of capitalism. And yet there is no denying Benjamin's profound intuition about the revolutionary temporality that laughter expresses.

According to Bakhtin's literary theory, the novel (organically adapted to new forms of silent perception, namely reading) inherits and develops the unaccomplished present, the actuality of time, the subjectivity that never coincides with itself. Let us add that the temporality found in cinema is a form of representation through time-images and that in television and digital media the present in the making, a time open to the future, is not only represented but constitutes the materials and the theme.

Benjamin was perfectly aware that responding to the industrialization of perception and the commercialization of the work by the reaffirmation of art was not only reactionary in the etymological sense but absolutely inefficient from a political perspective. The canonization of cinema as the seventh art is, for Benjamin, nothing but the other face of the Hollywoodization of the conditions of collective perception. It only reintroduced the respect and fear of the object that are nothing other than distance, the respect and fear of power. Benjamin warns us that those who gather in front of a work of art may at any moment assume religious behavior, reintroducing the absolute past and its ethics. Commercialization and art are the alternatives that power reproduces and that intellectuals must take on as a problem. The social division of time in contemporary capitalism can be described as follows: the present that passes in the cultural industry (the unfulfilled image interpreted only as perpetually disappearing, the present as simple repetition) and the past in art (the fulfilled image, time that remains and preserves). These are the recharacterizations that are subject to the emergence of power-time, that are simply reshuffled and prequalified temporalities of power as Bakhtin has described them.

The form of this splitting of time that calls for a recharacterization of the carnivalesque attitude toward technologies of time seems to find realization only in the diversion (*détournement*) that operates in the free, familiar contact of television, in the need to destroy distance and to approach the object of distraction and entertainment that the masses express anyway.

7.4

The concept of collective perception arouses other considerations that bring us back once again to Bergson and, more generally, to our present. The discontinuity of cinematic images, arranging themselves in continuous succession, produces aberrant movements in our perception, thus introducing us to the "optical unconscious, just as we discover the instinctual unconscious through psychoanalysis." For Benjamin, it is the cinema camera ("transformations, alterations, disasters of the visible world produced by the deformations of the camera") that allows access to the optical unconscious. The video camera, as we have seen, takes us even further in the discovery of pure perception (the Bergsonian unconscious). "For it is another nature which speaks to the camera rather than to the eye: 'other' above all in the sense that a space informed by human consciousness gives way to a space informed by the unconscious."[12]

Cinema produces a shock to the unconscious, enabling the masses to appropriate forms of perception of the psychopath, the hallucinator, and the dreamer, reversing the subordination of time to movement and including more reality than the perception of a healthy individual.[13] These new features of collective perception are, for Benjamin, a clear indication of a change in the function of the apparatus of human apperception. But "the tasks which face the human apparatus of perception at historical turning points cannot be performed solely by optical means—that is, by way of contemplation. They are mastered gradually—taking their cue from tactile reception—through habit."[14]

Here we find, surprisingly, another Bergsonian theme, a direct and unambiguous critique of the optical model. Vision without the passive syntheses of habit would be impossible. Any apparatus of vision needs its passive syntheses. The first objective of machines that crystallize time is not the eye (which, as we know, essentially functions as an extension of the intellect) but the body and action. First the body; the rest will follow. The reception of aberrant movement occurs through the tactile, and the shock of the succession of cinematic images introduces a "tactically dominant element . . . in optics itself."[15] And a distracted person might become accustomed better than anyone, because it is through the body and not the intellect that new images and temporalities are assimilated.

What the cinema announced is fully developed by video and especially by the computer, where we orient ourselves through a tactile optics, as we have come to realize. Recollection and contemplation block our familiarization with new technologies, because the production of perception is not primarily a fact of vision but of action. And the communications industry, though it demonizes real-time technologies and their images, familiarizes the humanity of tomorrow through games and habit. The automatic repetition involved in distraction and entertainment is at the heart of electronic gaming. And automation is one of the conditions for the development of the mind, since as Bergson says, it liberates the virtualities and possibilities of choice.

One of the main functions of art, according to Benjamin, is to familiarize ourselves with determined images, even before their ends have become conscious. If that has been partially accomplished in cinema, it has not even begun in video.

7.5

The shock produced by the cinematic image evokes the clash of workers with machines. Anticipating the Godardian relationship between the assembly line and the projection apparatus for edited images, Benjamin says that "first of all, with regard to continuity, it cannot be overlooked that the assembly line, which plays such a fundamental role in the process of production, is in a sense represented by the filmstrip in the process of consumption. Both came into being at roughly the same time. The social significance of the one cannot be fully understood without that of the other."[16]

But Benjamin establishes a different relationship between economic production and the cinema, since it no longer concerns the technological apparatus but the nature of the activity demanded of the spectator. Cinema technique, similar to sports, invokes the spectator's participation as a connoisseur or expert. Cinema (but also the press and sports) determines a cultural transformation in which the difference between the actor and the public tends to lose its unilateral nature. But this difference is only functional; it can vary from one case to another. "The reader is ready at any moment to become a writer."[17]

It is to Benjamin's credit that in the realization of this tendency, he connected it to transformations in work and to the rupture of the separation between manual and intellectual labor, of which cinematic production constitutes, according to him, the paradigmatic form. "As an expert—not perhaps in a discipline but perhaps in a post that he holds—he gains access to authorship. Work itself has its turn to speak. And its representation in words becomes a part of the ability that is needed for its exercise."[18]

The becoming-active of labor, the fact that it speaks, completely requalifies the role of art, because it reverses the basis of the social division of labor. Benjamin sees the performances of the dadaists, who oppose a distracted public to an artistic community that gathers together and contemplates, as an important symptom of the changing function of art. "With distraction, the artwork creates a shock which is nothing but a pretext for active behaviors of subjects."[19] The production and reception of art (but also of any work) can no longer be carried out independently of this second nature, its collective forms, technologies, and the active role played by the masses. The interactivity of digital technologies is based upon an underlying trend of behaviors and attitudes brought about by the mechanization of collective perception and distraction.

Critics, as already noted by Benjamin, rather than insisting on the specific ontological consistencies of new forms of perception and reception (the reversibility of the functions of the artist and the spectator, new processes of collective creation, etc.), bring us back, in the best cases, to art and, in the worst, to propaganda. Benjamin therefore sees in cinema the symptom of a radical transformation of the public, which not only becomes a mass-public by destroying bourgeois forms of perception but also acquires a new nature. Members of this mass-public, as experts who want to be actively involved as authors, are adequate subjects not only in perception but also in the process of the production of works. "Great works can no longer be regarded as the products of individuals; they have become a collective creation, a corpus so vast it can be assimilated only through miniaturization. In the final analysis, mechanical reproduction is a technique of diminution that helps people to achieve control over works of art—a control without whose aid they could no longer be used."[20] Collective forms of production and authorship, dissolving the separation between author and public, the active role of the spectator: these are the challenges posed by the conditions of collective perception to art, since it opens up the processes of singularization and the creation of a new nature. Has the cinema addressed these challenges? In any case, these problems have not even been posed in relation to video.

7.6

Adorno agrees with Benjamin about the need to separate kitsch cinema from cultural cinema. By contrast, he accuses Benjamin of not subjecting the two extremes of cultural production to the same dialectical treatment. "Both bear the scars of capitalism, both contain elements of change. They are the torn halves of freedom taken as a whole, which however is not achieved by adding the two."[21]

More generally, Adorno accuses Benjamin of underestimating the technical nature of autonomous art and overestimating that of dependent, commercial art. Without getting into the debate here, I would like to highlight the political theory that emerges from this analysis of cultural production. According to Adorno, Benjamin directly credited the proletariat with being the subject of cinema ("kino-subject"), a move he would not have made without a theory of intellectuals. Adorno refers directly to the Leninist theory of the party as a collective intellectual, as opposed to the blind faith that Benjamin puts in the process of the self-constitution of the proletariat within the historical process.

In my view, Adorno maintains a conception of the intellectual as avant-garde, while Benjamin sees in cinematic production a radical change in the figure and role of the intellectual. The reversibility of the functions of the author and the public anticipates the process of establishing an intellectuality of the masses that cinema announced and exponentially accelerated after 1968, resulting in the need to radically review the conditions of the revolutionary process. Spontaneity, action, and consciousness were completely displaced by the emergence of these new collective subjects and by the new reciprocal presupposition of perception and labor.

7.7

Collective perception, or the perception of the masses, should be the test of the revolution. If in advertising, art and perception as distraction are commercial evidence, in the revolution they will be human evidence. "If all conformed to cinematic capital, the process would stop the alienation of oneself, in the artist of the screen as well as within the audience."[22]

All analyses of Benjamin tend toward this key point: collective perception poses problems that can only be solved collectively. Revolution is, from this point of view, the attempt to innervate the collectivity with the organs that such technologies of mechanical reproduction create. What art should anticipate—"trends whose realization would have a destructive effect on the people themselves to win their right in the world of images"[23]—the revolution should realize in a collective form.

This process would be characterized by the disintegration of the proletariat as the masses and its constitution as a collective subject that alone can establish a harmony between the forces unleashed by technology and the human. The masses are the alienated form of the subject of collective perception. The tendency of the individual to break away from the masses, if it does not find expression in the revolution, will be exploited precisely in terms of the image, in the figure of the star and the reapplication of the religious functions of the cinema.

The revolution will not take place, and as Benjamin predicted, collective perception will be realized by the masses who will find, in the cameras of Leni Riefenstahl and Hollywood, a suitable eye: "In the great festive processions, in huge assemblies, in the collective organizations of sport and war—which are today offered to recording devices—the mass-public looks at itself through its own eyes."[24]

7.8

After Auschwitz—which ensured the mobilization of all the technical means of the time without putting into question their characteristics—the underlying tendency of collective perception, prophetically defined by Benjamin in cinema, is realized in another medium: television. Cinema no longer represents the conditions of collective perception, and any discourse that refuses to accept this development (virtually contained within cinema) is strictly reactionary. Actualizing the virtualities of collective perception contained in cinema creates a completely new situation that demands other tasks for art and for the collective element that must appropriate new virtualities created by the technologies of time. Wanting to fulfill these new tasks on the basis of the production and reception of cinematic images is not only illusory but condemned to be integrated with forms of expression of power.

Cinema has given us a second nature made of images. But the characteristics of this second nature (the optical unconscious, omnipresence, the explosion of the world through the dynamite of tenths of a second) are only represented. Cinema shows us movement and time. And it can make us see all its syntheses, because it works with duration-images. But this representation-vision always takes place in a deferred time. Cinema, because of its technology (the separation of the shot and its dissemination, or, according to a suggestion made by Sergei Eisenstein, the separation of shooting and editing), still retains the distinction between reality and image, between the actual and the virtual. Television brought us to another dimension where these distinctions no longer apply. The fundamental reason for the change is the fact that television, operating in real time, doubles the world with its images, covers it with a layer of image memories, at the very moment something happens. As we have seen, its essence is to be internal to time in two forms: internal to time-matter, where it contracts and dilates vibrations, and internal to pure memory, where it preserves itself while splitting every moment into a present in the making and a movement into the future. With television we have entered the spectacle, in which there is an indistinction between the thing and the image, the real and the imaginary, the actual and the virtual, given their continuous exchange.

With cinema we are in the dimension of shock (the predominant sensation of the big industry era), but with television we are in the dimension of flow. Cinematic images shock because they open the world of the optical unconscious to a space and a time beyond human experience, a Bergsonian world composed solely of images, but with a distinction between the real and appearance, between the actual and the virtual. The magic screening room, a place of celebration for the cultic worship rendered by this new world, alone had the power to momentarily keep us prisoners of this illusion. But now flows entirely envelop us: *noi andiamo in onda* (literally, "we're going on air"), as perfectly expressed in Italian. Not only television programming, but all of reality is captured in this movement to go on air. The image no longer shocks us because it is no longer external to our perception; we ourselves have become images. Only television can achieve the indistinction between the actual and the virtual, as announced by cinema. Cinema introduced movement and time in the sequence of images (shock), but television is the same movement of time-matter (flows) and its modulation.

If cinema has generalized the value of exhibiting art by ratcheting it down ad infinitum while retaining the public place of cultic worship, television deterritorializes the place of this worship in any space whatsoever, thus eliminating the value of exhibiting. What is exhibited is the same indistinction of the world and the image. Television reclassifies the differences between space and time, public and private, individual and collective, based on a Bergsonian nonchronological time.

After Auschwitz, television destroyed the public. The socialization of perception and the individualization of reception go together. Networks will complete the destruction of the public, in the sense that they introduce a reversibility between author and public, between production and consumption, thus making these functions highly productive. Reception functions as distraction because it no longer takes place where it collects itself; in other words, distraction has become the very form of perception. In any case, what is attention to the image when it is not distinct from the object it describes?

Postwar cinema perfectly represented and anticipated this new dimension, showing us a direct image of time in which we can no longer distinguish the actual from the virtual. But with television, there is no longer representation, because television is itself a direct image of time. Video is time. Cinema is but one symptom of this new dimension. Cinema is an adventure of perception, whereas television is an adventure of time.

7.9

Television raises other problems: it is no longer only about images that represent, but images that are genetically constitutive of the world. The representation-image imposed upon us by television is an apparatus of power. It is therefore useless to search exclusively for representation-images in video, because for images it is necessary to make things, to construct situations, events, forms of life. Insisting on the visibility (or better the nonvisibility) of video images is a false issue that brings us back to cinema. We must recover the adventure of perception that has been a crucial experience for humanity, but in order to insert it into this new dimension. And insert here means creating something new, even in cinema. As we know, the video image is a tactile image, an image in which to intervene rather than to watch.

The conditions of tactile perception-reception refer, according to Benjamin, to the experience of architecture rather than that of painting, in which familiarity arises from habit rather than contemplation. Or we could, from what we have tried to demonstrate about television, speak about temporal architecture: how to inhabit time, familiarize ourselves with the new temporalities, and from these new habits construct other space-time dimensions.

The video apparatus is used not only to see (as its etymological root would suggest) but to create situations, to intervene in the event. It requires a response; it implies the activity of the spectator without which, as Nam June Paik says, it cannot come into existence. And indeed, what existed then was television and not video. The passivity that constrains us by the power of the television apparatus is directly proportional to the activity that the ontological consistencies of video elicit: the image in the making, the situation in the making, the subjectivity in the making; in short, nonchronological time.

Furthermore, all the ontological consistencies of video and the activities of the spectator will reappear, inevitably, with computer and digital networks. Passivity with activity, perceptual isolation with the hypercommunication of all with all, the separation between production and reception with their most thorough integration. The visibility of the image will integrate the operativity of the human-computer couple. We are not visionaries but actants.

7.10

The passive-visual usage of the television spectator, the reduction of all the virtualities of television to an instrument of unilateral reception administered according to the neutralization of the event, must be explained by the regime of temporality that dominated Fordism: the subordination of power-time to metered time (of value). It is this temporality that controlled and still controls the ability to produce and reproduce the real time of television. All the ontological consistencies of video are selected and subordinated to metered time and its organization.[25]

The emergence of other social temporalities (after 1968) has revealed other virtualities of technological apparatuses that will develop beyond television in another medium: digital.[26] Benjamin continues here to guide us in these passages. He notes that cinema and Taylorism (the assembly line and the chain of assembled images) are nearly contemporary. Taylorism is interpreted by Benjamin as a process that deprives the worker of experience; craft, know-how, cooperation, and power constitute this experience. Labor is reduced to a series of movements to be performed (shocks) according to orders. The worker must not act but react. The consumer, illustrated by Benjamin through the attitude of the player, is subject to the same relations of stimulus-response. Fordism attempts to reduce humans and their activity in the sensory-motor schemas[27] to the docile body of the Foucauldian factory.

For Benjamin, the worker is now subjected to the "test": carrying out movements codified under the gaze of experts (the office of management and planning) or of a machine. From this point of view, cinema is the experience of the exam that, always before a machine, reproduces and measures the actions and behaviors of the masses.[28]

But "work speaks," as Benjamin says but above all refuses. And he refuses the division between manual and intellectual labor; he revolts against the separation and the expropriation of intellectual, communicative, and linguistic functions and their attenuation in sensory-motor schemas. The refusal of work is the refusal of this condition, a refusal of work that could be interpreted, in relation to time, as the refusal of the splitting of the whatever-time of capitalism into metered time and power-time and the subordination of the latter to the former. The great development of television and digital networks corresponds to the moment when this refusal was completely achieved (after 1968), when time liberated from any measure appeared as the source of production as whatever-time, beyond the division between the time of labor and the time of life.

The emergence of power-time requalifies and redistributes in a new way (in relation to time) all divisions of capitalism. The emergence of power-time[29] also requalifies the indistinction of the actual and the virtual (and their circuit), which television showed that it operates at the social level. The indistinction between the real and the imaginary, between the image and the thing, was, under the primacy of metered time, to block and neutralize their power of creation. The actual-virtual circuit of television, subordinated to Fordist temporality, functioned as a new closing of time. No longer the perfect circle of the absolute past but the enchanted circle of the infinite return, of the sterile reflection of the image and the thing. But the emergence of power-time, the rupture of the subordination to metered time, interrupts the enchantment of this bad infinite, interrupts the crystal of the continual reflection of the actual and the virtual, and realizes the conditions from which this circuit becomes the source of unpredictable new creations. Digital apparatuses are the technological translation of this passage because they render productive the actual-virtual circuit and constitute the technological conditions that make it possible to escape the vicious circle of their presuppositions and their reciprocal contemplations. They redefine, on the basis of a temporal monism, the differences between matter and spirit, between subjective and objective, between time and space.

Digital apparatuses are not limited to doubling the world with images (as in television) but are the source of a new sensible and intelligible and establish a new materiality and spirituality. They emphasize that time is at the origin of subjectivity and matter, of production and creation, and that their differences are nothing other than modulations, solidifications, and repetitions of time. A new power of metamorphosis and creation is at our disposal. New forms of subjectivity and materiality are now possible. The world is time; these technologies do not interpret it according to the uniformity of value, but under the continual possibility of creation that the constitution of nonchronological time carries with it.

The fortune of the postmoderns, their ideological task, consisted in this: release the sterility of the actual-virtual circuit at the very moment it was beginning to show its full power. They discovered the spectacle precisely when we entered another dimension, and they did nothing but celebrate it. Rather than indicate a new ground of confrontation, along with its new set of problems, they have simply seduced and fascinated with their theories about the disappearance of the world. But now the situation is radically different. There is no longer, as in Benjamin and Fordism, a technological apparatus for collective production and perception. But it is one and the same apparatus—digital technology—we perceive and work with, whose raw material is not the time of labor but time as such. The separation between production and reception is blurred, since the same apparatus can do both things at once. If these divisions remain (and they do), they have only one functional and political ground. All the qualifications of collective perception that we find in Benjamin are here actualized on the basis of the power-time that requalifies them in the direction of creation and activity.

Let us dispel any ambiguities and objections that this theory might bring up. That there are no more distinctions does not mean that we have entered the undifferentiated. We just need another ground for defining the differences, a temporal ground. Intellectual and manual labor, the time of work and the time of life, the image and the thing, the real and the imaginary, time and space, do not disappear but receive another qualification by the emergence of power-time. This nonchronological time distributes them within a new nature, which makes them reversible, less rigid, more flexible; a time that appears directly as the source, the origin of functional and nonobjective differences. A new conception of ethics should be the basis for their differentiation, because the degree of freedom that they permit is increased, as we have seen with Bergson (liberation from the cursed necessity of labor, to speak in Marxian terms). Therefore, I am simply attempting to describe the ontological consistencies of the new conditions of perception-production. This does not mean that new divisions will not be able to occur (they already have), but that these divisions, on the basis of a new nature, can only come back to ethics or power.

7.11

How to regain singularity and escape from the indistinction between the actual and the virtual, the reversibility of material and immaterial labor, the subjugated reversibility of capitalist accumulation between the time of work and the time of life? How to render their relationship as destructive-creative? The real has not disappeared, the social is not already given, but must be crystallized each time. The real and the social need to be continually created, reinvented. The machines that crystallize time play a strategic role, because within the indetermination of this always unaccomplished time, they embody the technological conditions of the coproduction of the real and subjectivity. Both the real and subjectivity find in machines that crystallize time a new power of metamorphosis, of modulation, of creation. Here the conditions of perception and labor, in their mutual reference and reciprocal presupposition, are the conditions of the cocreation of the world.

The power that the actual-virtual circuit expresses, once liberated from its subordination to metered time, must determine the processes of singularization and reterritorialization that escape the information economy. And aesthetic assemblages, with their force of singularization that is always inventing new worlds, may become the paradigm upon which to measure this new production. But these assemblages should be verified and confronted with the new conditions of collective perception and labor as well as their indistinction-reversibility.

To verify and confront means to create apparatuses that make it possible that individual or collective bodies are in a position to emerge as new existential territories. Only the regulation by collective assemblages in the production of subjectivity makes it possible to invent singular assemblages. Benjamin's assertion of the need to collectively solve the challenges presented by the socialization of perception and labor is taken up by Guattari. But here the collective, to the extent that it intensifies and enriches these faculties (including time in its own constituent fabric), is singularized and individualized. The mass-public has exploded into minorities and can no longer find in the concept of a general and totalizing class its human verification. The collective element that was to verify the revolution-disintegration of the masses and the public went beyond Benjamin's wishes.

The transformation of the functions of art, widely anticipated by video and further expressed by digital technologies, is summed up in Guattari's formula, which states that art should not only tell stories but create apparatuses in which the story can exist. Aesthetic practices thus become highly productive, as verified in the information economy, because here too the distinctions between art and life, between art and work, tend to lose their unilateral character, as Benjamin had announced.

Therefore we end as we began, hoping for the emergence of a new type of barbarism in which power-time opens an incommensurable field of action with the time that has been lost. Friedrich Nietzsche, in the quote highlighted in the introduction, saw in the crisis of the socialist regimes constituted on the project of becoming-masses, becoming-proletariat—and that fell with the Berlin Wall—one of the conditions of its appearance. Benjamin reminds us that the power-time in which we are living is another condition for the creation of a new barbarism.

> Barbarism? Yes, indeed. We say this in order to introduce a new, positive concept of barbarism. For what does poverty of experience do for the barbarian? It forces him to start from scratch; to make a new start.[30]

> He sees nothing permanent. But for this very reason he sees ways everywhere. Where others encounter walls or mountains, there, too, he sees a way. But because he sees a way everywhere, he has to clear things from it everywhere. Not always by brute force; sometimes by the most refined. Because he sees ways everywhere, he always stands at a crossroads. No moment can know what the next will bring. What exists he reduces to rubble—not for the sake of the rubble, but for that of the way leading through it.[31]

AFTERWORD

Videophilosophy Now—an Interview with
Maurizio Lazzarato

JAY HETRICK: Your work is increasingly getting published and recognized in English translation; for example, *The Making of the Indebted Man* appeared recently with Semiotext(e), and soon *Experimental Politics* will be published by MIT Press. However, English-language readers still lack a reliable biography of you. Could you please give us a brief synopsis, especially in relation to how you conceived of *Videophilosophy*?[1]

MAURIZIO LAZZARATO: Next spring another book, *Signs and Machines*, will appear in English with Semiotext(e), and yes, after that *Experimental Politics* will be translated. Soon nearly everything I have written will be available in English, except the book on Gabriele Tarde. Following *Signs and Machines*, Semiotext(e) will publish a short book on Marcel Duchamp and the refusal of work—it is nearly completed. In any case, my biography is very simple, but I will not start when I was a small boy! I went to France for political reasons in 1982 because I was having problems in Italy, and there I finished my studies. I have lived and worked in Paris as a more or less independent researcher ever since, albeit associated at different times with the Centre national de la recherche scientifique, l'université Paris 1, and the Collège international de philosophie. I did my DEA at the École des hautes études en sciences sociales and then went to Saint Denis (l'université Paris 8), where I finished my doctorate on video philosophy in 1996. Of course, I originally wrote it in French but then rewrote and published it in Italian.

JH: How does *Videophilosophy* fit within your overall intellectual trajectory?

ML: There is a very strong connection between this book and the recent collaborations I have done with the video artist Angela Melitopoulos, particularly our video installation *Assemblages*.[2] Twenty years ago the main problem was how to escape the subject-object relation, how to problematize the paradigm of modern subjectivity. This is also what we see in *Assemblages* in terms of content and, perhaps, aesthetically as well. In *Videophilosophy* I tried to get out of this subject-object relation through the image, through Henri Bergson's ontology of the image—especially as he presents it in the first chapters of *Matter and Memory*, which is an extraordinary text in the history of philosophy—where he claims that everything is image, the real is image. There is no separation between a subject that sees what we call an "object" and the object itself; both are images. It was a philosophical ordeal to get out of that modern opposition between subject and object. But we must realize that Bergson's image is not the kind of image we normally think of. The image that Bergson is talking about when he refers to pure perception is more like vibrations. Video technology functions through and on that pure perception of vibration. It functions as a synthesis and as a memory. It is fundamental to see that video is radically different from cinema. Recently, I was working on Pier Paolo Pasolini and realized that he has constructed an immanent semiotics that contains a concept of the image very close to Bergson's concept. But video is really the world reduced to a flow. The main thing is to see that video is a means to move beyond the opposition between subject and object—which actually reduces the multiplicity of the world—or it is a way of reaffirming this multiplicity beyond television. In any case, looking back twenty years, I realize that all the themes that I am working on now were already in *Videophilosophy*. I see all the ideas, such as asignifying semiotics and the problem of subjectivity, but since then our political problems have become more urgent, the economic crisis has become stronger, and the situation has become more traumatic, so my tendency recently has been to consider the possibilities for intervention more on the level of politics. But the philosophical foundation remains the same. So, for example, in *Videophilosophy* there is a particular reading of Bergson and Friedrich Nietzsche, which is now oriented politically.

JH: In the introduction to *Videophilosophy* you pose Nietzsche and Bergson contra Karl Marx, making the bold claim that "while Marx indicated the methodology with which to discover 'living labor' beyond work, he is of no help in discovering the forces" that underlie the conditions of contemporary capital. You add that it is, in fact, Nietzsche and Bergson who "should be understood as the conceptual personae who have constructed an ontology" adequate to the logics of post-Fordism and immaterial labor.[3] So is it fair to say that you are doing two things with this book: constructing a video philosophy on the one hand, and on the other, constructing a political ontology that might serve to ground your more explicitly political work?

ML: Yes.

JH: After all, *Lavoro immateriale* and *Videofilosofia* were both published in Italy at the same time, in 1997, with the latter being a more "philosophical" work.

ML: That was twenty years ago, so today I might articulate these things a bit differently. With Nietzsche and Bergson, we are able to grasp the complexities of subjectivity and the micropolitical level of politics more generally. Nietzsche and Bergson go further with regard to the problem of subjectivity, which for me is crucial for understanding the dynamics of contemporary capitalism. But of course, Marx is still important for understanding the nature of capital itself. The problem with Marx is that he was thinking solely within the paradigm of subject-object relations. Subject-object relations still exist and somehow function, but there is a deeper problem. On the one hand, there is the level of subject-object relations, the paradigm that relates given subjects on a macro level. But on the other hand, there is another function of subjectivation that I call, after Félix Guattari, machinic enslavement. On this level you cannot really deal with Marx but must pass through thinkers like Bergson and Nietzsche. Regarding the concept of immaterial labor: at that time it was not very clear, so I took a more experimental approach, but perhaps today it is still not so clear. In any case, going beyond the subject-object relation means departing from representation. This means representation in language, but also in politics. So what does it mean to make a film, or construct a politics, without representation? In the 1970s we saw several theories of semiotics that tried to move beyond representation, one of them Pasolini's, in which nonverbal signs are prior to language. Pasolini understood the transformation of this semiotics in relation to the transformation of capitalism and therefore used Marxist terminology to describe it. It is not the superstructure but the base that is more important. So with him we have the language of production and consumption rather than institutions and schools. Fernand Deligny offers another way to escape representation: what can we do when there is no language? The semiotic theory of Gilles Deleuze and Guattari is also important, though it is not always recognized as such. In it we have several different semiotic registers that connect to or engage with one another: language, symbols, and gestures, as well as asignifying semiotics like money or informatics. These signs are all productive without entering consciousness.

JH: I was asking about political ontology because we find, roughly speaking, a historical split within the Left between orthodox Marxism and various forms of non-Marxist socialism and anarchism. Of course, in the mid-nineteenth century Friedrich Engels described this as a split between scientific (i.e., dialectical) and utopian socialism. One strain of the latter is marked by a supposed "spontaneism" that occasionally seems to resonate with certain ideas of Georges Sorel, Antonio Gramsci, and perhaps also Walter Benjamin, all of whom have some degree of connection to Bergson. After Bergsonism was completely, and perhaps unjustly, dismissed by critical theory, other approaches attempted, in varying degrees, to move beyond or around G. W. F. Hegel through the work of Baruch Spinoza, let's say in the work of Louis Althusser, Antonio Negri, and others. How do you see your work in relation to this broad history? Do you think it is necessary—following Deleuze's construction of a minor tradition of metaphysics—to construct a minor, non-Hegelian tradition of critical and political theory that might be more relevant for postautonomous thought?

ML: Perhaps, but when I wrote *Videophilosophy*, my direct references came from French philosophy of the 1960s and '70s, especially Deleuze, Guattari, and Michel Foucault. My reading of Bergson, for example, follows from that of Deleuze rather than from a minor history of the Left. But yes, today we see another moment of cross-pollination between anarchism and communism. We could say that the communism discussed in the nineteenth century would never work today without being supplemented with certain anarchical forms of organization.

It is, in fact, precisely these forms of organization that have to be reinvented on a political level today. But on a theoretical level, we must keep the references to recent French philosophy. As for using Bergson or Spinoza contra Hegel, I'm interested in Bergson because with him it is possible to theorize the technology of image production in a way that is not really possible with Spinoza. Furthermore, while Spinoza might be useful for constructing an ontology, for me it would be problematic to use Spinoza to analyze the contemporary conditions of capitalism.

JH: Getting back to *Videophilosophy*, we see that Bergson offers a novel ontology of images that implies a form of experience beyond the strictures applied to it by reason. In Bergson's project experience is extended through a theory of perception deliberately pushed beyond the usual Kantian coordinates. In a few of your own writings you pick up this idea of expanded experience through Benjamin, but also through William James. Bergson's philosophy, especially as Deleuze reads it, presents an onto-aesthetics that, you argue, video adequately expresses. But since György Lukács, the concept of expanded experience has been all but abandoned as a reactionary notion linked to vitalism. I wonder, however, if such a concept is indeed needed for us to think the event: we are not yet seeing the situation clearly. You quote Benjamin at the end of *Videophilosophy*, stating that in order to construct a "new barbarism," we have to first testify to our relative poverty of experience within the new conditions of capital. So again, we move from video to politics.

ML: You could say that, but we also have to understand the medium in which we are working. With a book or with video art like that of Angela Melitopoulos—who I've been working with for twenty years—there is the question of praxis. It is easy to understand the expansion of experience through the use of video technology. We expand the field of vision with the use of technology. Coming back to what Bergson said, we know the Kantian subject sees less than what exists in reality. If we want to understand the real, we have to take leave of the Kantian subject. This can be done with the help of James or Bergson or with the help of video without James or Bergson! Of course, this is also a problem of everyday life. Political change entails not simply asking for better work, better living conditions, and so on but also widening one's field of experience. Many problems of subjectivity related to work are precisely connected to this idea of expanding experience. So the crisis we are living in today is not just an economic crisis but also a crisis of the production of subjectivity. Therefore, in my work, thinking about the production of subjectivity is fundamental. This is what you don't really find in Marx, or even in the Marxist tradition. But in the tradition of James, Bergson, and Nietzsche, there are many ways to think through what this production of subjectivity means. When we say there is a lack of real political organization, we might have to first discuss the necessity of widening the field of experience. The Benjamin quote about a "new barbarism" is still very actual today, since the poverty of experience points to the necessity of having new barbarians rise up. In *A Thousand Plateaus* Deleuze and Guattari discuss the barbarian in relation to the nomad, who crosses borders continuously, who enters into and exits the empire freely.

JH: But who are the new barbarians today? The multitude—which Thomas Hobbes characterized as an "insolent rabble"—or something more specific? And how does it relate, if at all, to the concept of animism that you borrow from Guattari and use as a conceptual trope in *Assemblages* with Melitopoulos?

ML: If you go to Brazil and witness the different political manifestations and demonstrations, it is obvious that a new form of barbarism is emerging. From the perspective of the Workers'

Party—they see what is happening—it's as if they are intruding from the outside. And those demonstrating at Gezi Park against Turkish prime minister Recep Tayyip Erdoğan are barbarians. They come from inside the country, but they are barbarians. As for animism, Melitopoulos proposed the idea, and then I started researching the many references on animism in Guattari's work. This is related to the problem of how to exit from the Kantian subject, the subject of modernity. And the contemporary situation is very telling. At the end of his life Guattari argued that contemporary society had to go through a new phase of animism. However, this did not entail going back to an archaic form of animism, but rather through a new way of thinking in order to find a new relationship between subject and world. Western modernism has emptied the world of spirits and focused its efforts instead on the construction of one kind of subject. Since Kant, we have had the human at the center with everything else revolving around it. Deleuze and Guattari try to decenter the subject by posing the problem of subjectivation. Subjectivity, as Eduardo Viveiros de Castro says, is not only, not necessarily, human subjectivity. We can find it in machines, in objects, in everything that surrounds us. We are not the only ones able to act; having agency is not just a human capacity. To think this requires a new animism—a machinic animism—is very contemporary, is what is happening to us now, as Guattari suggests. This way of thinking allows us to escape from the subject, but also from monotheism and all other ontologies of the one.

JH: Guattari's aesthetic paradigm, you claim, should be understood as referring to aesthetics pure and simple; that is, to *aisthesis*, sensation, and more specifically, to an onto-aesthetics of asignifying signs. And in a way, this brings us back to the expanded perception we were talking about earlier. Maybe we could modify Jacques Rancière's phrase and talk about the redistribution and expansion of *sensation*. But as both you and Guattari note, the aesthetic paradigm also entails a new concept of creation, which extends beyond the artwork to include the real itself—as in Bergson's metaphysics—as well as to ethics and to the construction of the social. But isn't the concept of creation, perhaps like the idea of an extended experience beyond Kant, a bit too mired in romanticism and a kind of thinking that some may deem uncritical? At the same time, creativity is already inherent to the logic of contemporary capitalism and the type of "creative" subject it wants to create: *You*Tube, *i*Phone, *Just Do It*. So how do we know we are on the right track with this new concept of creativity?

ML: We can understand the limits of Marxism from the perspective of the aesthetic paradigm. Marxism still functions within a scientific paradigm. Guattari says that we have to abandon the scientific paradigm, since science itself can be explained through creativity. We can find a tradition—especially at the end of the nineteenth and the beginning of the twentieth century—of the aesthetic paradigm, for example, in the works of Benjamin, Bergson, James, and Nietzsche. Nietzsche says we are all artists; we understand ourselves as artists. Even in the most banal perception, we are always creating, always adding something. But now we have to find a political form adequate to this aesthetic paradigm. Because creativity concerns not only the arts but also the organization of society, we have to find a corresponding politics to articulate it. So there is an important political transformation that must begin with the aesthetic paradigm. Of course, contemporary capitalism has appropriated the idea of creativity (e.g., in Nike slogans), but this creativity does not allow for the expansion of experience. There is the production of a kind of change, a production of the new, but it functions in a way that impoverishes the field of experience. Guattari says that today there is no real creativity anywhere. In business, in particular, there is no creativity. For him, the word *creativity* points to something very specific: the production of singularity. Duchamp discusses creativity in relation to the modern social division of work: this is a doctor, this is an artist, and so on. Even then, but especially now, art is an institution that simply functions within the capitalist system. Duchamp argues that as long as we work within these divisions, we will never come to an understanding of what creativity really is. He sees this and refuses to be labeled an artist, since the very carving out of a specific type of "artistic" work within the social field renders creativity impossible. I think this is insightful. The interesting thing about Duchamp is that he tried to place himself on the border between art and nonart. It is a difficult position to assume.

JH: Should the aesthetic paradigm as developed by Guattari and taken up in your own work be regarded as a model for a new type of political action? Or is the construction of an ethics sufficient? You talked about the construction of new subjectivities as being necessary, but is it simply a first step? How can we move from this toward the construction of new political forms adequate to the contemporary conditions of capital? Are we still somehow caught up, paralyzed perhaps, in this moment of negotiating the problem of subjectivation? And is this one reason why we have seen very little in terms of real, viable political programs offered by the Left: we're still floundering at the level of constructing new forms of subjectivity that function to counter all forms of fascism—to speak with Foucault—especially on the micropolitical level? In her book *Capitalist Sorcery*, Isabelle Stengers argues for the necessity not of constructing a political program but of simply developing an ethics of the scream that testifies to the events in Seattle (the 1999 World Trade Organization protests).

ML: Yes, this is the situation in which we find ourselves today. We see a series of struggles arising post-Seattle, more radical struggles posing real problems. You see it in Brazil and Turkey, but in all these struggles we haven't yet found an effective mode of organization and therefore don't understand how to oppose power. Historically, the working-class movement figured out how to strike and resist capitalist aggression. We have to construct something new on the social level, understand the multiplicity of the composition of contemporary society. It is not simply the working class anymore, but a multiplicity of subjects. And then we have to think about political organization, of course. But we are in a period of austerity. The five hundred wealthiest individuals in France have a higher net worth than 25 percent of the population. We do not know how to reverse this kind of inequality. We have to figure it out, but at the moment we do see new forms of subjectivation, new fields of experimentation, emerging everywhere. Maybe we need some more time. But of course, I think that Deleuze, Guattari, and Foucault are absolutely crucial for thinking through these problems, more interesting and relevant than the ideas put forth by Giorgio Agamben, Alan Badiou, and Rancière.

July 2013
Amsterdam

NOTES

LAZZARATO'S POLITICAL ONTO-AESTHETICS

1. The present volume is based on a revised but unpublished French version of the text, though I have also consulted Lazzarato's original thesis, the published Italian version (*Videofilosofia: La percezione del tempo nel postfordismo* [Rome: Manifestolibri, 1997]), and the German translation (*Videophilosophie: Zeitwahrnehmung im Postfordismus* [Berlin: b-books, 2002]).
2. Introduction in this volume.
3. An excellent analysis of Lazzataro's political philosophy, especially in relation to his work on Tarde, can be found in James Muldoon, "Lazzarato and the Micro-politics of Invention," *Theory, Culture and Society* 31, no. 6 (2014): 57–76.
4. Maurizio Lazzarato, "Debt, Neoliberalism, and Crisis," *Sociology* 48, no. 5 (2014): 1043.
5. Chapter 6 in this volume.
6. Maurizio Lazzarato, "Grasping the Political in the Event," *Inflexions* 3 (October 2009): 14.
7. See, for example, Alberto Toscano, "Vital Strategies," *Theory, Culture and Society* 24, no. 6 (2007): 71.
8. Lazzarato clearly prefers Foucault's micropolitics to Gramsci's political philosophy. Maurizio Lazzarato, *The Making of the Indebted Man: An Essay on the Neoliberal Condition*, trans. Joshua David Jordan (Los Angeles: Semiotext(e), 2012), 107.
9. François Azouvi, *La glorie de Bergson* (Paris: Gallimard, 2007); Carl Schmitt, *The Crisis of Parliamentary Democracy* (Cambridge: MIT Press, 1988).
10. Work in this direction has fortunately already begun. See, for example, Monica Greco, "On the Vitality of Vitalism," *Theory, Culture and Society* 22, no. 1 (2005): 15–27.
11. By placing discontinuity at the very heart of his cinematic metaphysics, Deleuze may be responding to Gaston Bachelard's critique of Bergson in

his *Dialectic of Duration*. "Since any critique is clarified by its end-point, let us say straight away that of Bergsonism we accept everything but continuity. . . . We wish therefore to develop a discontinuous Bergsonism." Gaston Bachelard, *The Dialectic of Duration*, trans. Mary McAllester Jones (London: Rowman and Littlefield, 2016), 20.

12. Anne Sauvagnargues, *Artmachines: Deleuze, Guattari, Simondon*, trans. Suzanne Verderber (Edinburgh: Edinburgh University Press, 2016), 55.
13. Antonio Negri, *Time for Revolution* (London: Continuum, 2003); Éric Alliez, *Capital Times* (Minneapolis: University of Minnesota Press, 1996).
14. "With the end of the gold standard declared by Nixon, the year 1971 represents the outset of this history." Lazzarato, "Debt, Neoliberalism, and Crisis," 1043.
15. Introduction in this volume.
16. Chapter 6 and introduction in this volume.
17. Chapter 2 in this volume.

The definition of machines that crystallize time will enable us to put intellectual labor at the center of the production of images and syntheses of time because seeing, listening, understanding, interpreting, and creating are produced by the activity of synthesis of intellectual labor. But it must be emphasized immediately that Bergson's concept of intellectual labor does not refer to the intellectual/manual labor couple that Marx tells us lies at the foundation of all capitalist divisions of labor. The Bergsonian concept falls short of the divisions of labor, but can be very useful in understanding the nature of post-Fordist labor. (Maurizio Lazzarato, "Les machines à cristalliser le temps" [thesis, University of Paris 8, 1996], 43, http://octaviana.fr/document/17455107X)

18. Ernst Bloch, *The Heritage of Our Times*, trans. Neville and Stephen Plaice (Cambridge: Polity, 2009), 322.
19. Toscano "Vital Strategies," 86.
20. Chapter 2 in this volume.
21. Axel Honneth, "A Communicative Disclosure of the Past: On the Relation Between Anthropology and Philosophy of History in Walter Benjamin," *New Formations* 20 (Summer 1993): 86.
22. Chapter 7 in this volume.
23. Walter Benjamin, "On the Concept of History," in *Selected Writings*, vol. 4, *1938–1940*, ed. Howard Eiland and Michael W. Jennings (Cambridge: Harvard University Press, 2003), 396, 391.
24. Peter Hallward, *Out of This World: Deleuze and the Philosophy of Creation* (London: Verso, 2006).
25. Maurizio Lazzarato, *Marcel Duchamp and the Refusal of Work*, trans. Joshua David Jordan (Los Angeles: Semiotext(e), 2014).

26. Gilles Deleuze, *Cinema 1: The Movement-Image*, trans. Hugh Tomlinson and Barbara Habberjam (Minneapolis: University of Minnesota Press, 1986), 83.
27. "The camera liberates perception and thought as well as the center of gravity that defines the human body. The kino-eye, which Vertov also called the machine-eye, moves in a perpetual metamorphosis—a discontinuous movement of bodies." Chapter 1 in this volume.
28. Chapter 1 in this volume.
29. Chapter 1 in this volume.
30. Chapter 1 in this volume.
31. Maurizio Lazzarato, "Existing Language, Semiotic Systems, and the Production of Subjectivity in Félix Guattari," in *Cognitive Architecture: From Bio-politics to Noo-politics*, ed. Deborah Hauptmann and Warren Neidich (Rotterdam: 010, 2010), 512.
32. Those who want to explore other ways in which Bergson's philosophy may ground a metaphysics of new media can consult Stephen Crocker, *Bergson and the Metaphysics of Media* (Basingstoke: Palgrave, 2013).
33. Chapter 3 of this volume.
34. Chapter 3 of this volume.
35. Lazzarato, "Existing Language," 515.
36. Lazzarato.
37. Chapter 3 of this volume.
38. Félix Guattari, "On Machines," *Journal of Philosophy and the Visual Arts* 6 (1995): 9.
39. Guattari.
40. Guattari, 8.
41. Maurizio Lazzarato, "The Machine," trans. Mary O'Neill, European Institute for Progressive Cultural Policies, October 2006, http://eipcp.net/transversal/1106/lazzarato/en.
42. For an analysis of this work, see Jay Hetrick, "Video Assemblages: 'Machinic Animism' and 'Asignifying Semiotics' in the Work of Melitopoulos and Lazzarato," *Footprint* 14 (Spring 2014): 53–68.
43. Maurizio Lazzarato, "Semiotic Pluralism and the New Government of Signs: Homage to Félix Guattari," trans. Mary O'Neill, European Institute for Progressive Cultural Policies, June 2006, http://eipcp.net/transversal/0107/lazzarato/en.
44. Félix Guattari, *Molecular Revolution: Psychiatry and Politics*, trans. Rosemary Sheed (London: Penguin, 1984), 127.
45. Lazzarato, "Semiotic Pluralism."
46. Introduction in this volume.
47. Lazzarato, "The Machine."
48. Antonio Negri, *Art and Multitude*, trans. Ed Emery (Cambridge: Polity, 2011), 29.

49. Negri, 26, 31.
50. Maurizio Lazzarato, "The Aesthetic Paradigm," in *Deleuze, Guattari and the Production of the New*, ed. Simon O'Sullivan and Stephen Zepke (London: Continuum, 2008), 175.

INTRODUCTION

1. Henri Bergson, "The Perception of Change," in *The Creative Mind: An Introduction to Metaphysics*, trans. Mabelle Andison (New York: Dover, 2007), 109.
2. "Machines which run on oil or coal or 'white coal,' and which convert into motion a potential energy stored up for millions of years, have actually imparted to our organism an extension so vast, have endowed it with a power so mighty, so out of proportion to the size and strength of that organism, that surely none of all this was foreseen in this structural plan of our species: here was a unique stroke of luck, the greatest material success of man on the planet." Henri Bergson, *The Two Sources of Morality and Religion*, trans. Ashley Audra and Cloudesley Brereton (Notre Dame: University of Notre Dame Press, 1977), 309.
3. Bergson, 312. Translation modified.
4. "For we feel that a divinely creative will or thought is too full of itself, in the immensity of its reality, to have the slightest idea of a lack of order or lack of being. To imagine the possibility of absolute disorder, all the more the possibility of nothingness, would be to say that it might have not existed at all, and that would be a weakness incompatible with its nature which is force." Bergson, *The Creative Mind*, 47.
5. Nietzsche, *The Will to Power*, trans. Walter Kaufman and R. J. Hollingdale (New York: Vintage, 1968), 421.
6. Henri Bergson, "The Possible and the Real," in *The Creative Mind*, 73.
7. "Nietzsche's metaphysical thesis, namely that all things are mere appearance, abolishes the difference between being and appearance, which Hegel could not save through the negative, and in raising this difference, appearance now reflects essence. . . . However, this metaphysical assertion, although it has been abused ideologically, disguises a social content: it expresses alienation, which is objectified in an immediate and non-conceptual totality. Negative ontology, which abolishes all reality by reducing it to appearance, reifies this appearance itself, turning it into being. Nevertheless, whether we like it or not, it registers a social experience." H. J. Krahl, *Costituzione e lotta di classe* (Milan: Jaca Book, 1973), 137.
8. Nietzsche, *The Will to Power*, 268. Translation modified.
9. Bergson, *The Creative Mind*, 2. Translation modified.
10. Bill Viola, "Y aura-t-il copropriété dans l'espace de données?," in "Vidéo," ed. Raymond Bellour and Anne-Marie Duguet, special issue, *Communications* 48 (1988): 72.

11. Nam June Paik, interview by Angela Melitopoulos, *Mixed Pixels: Students of Paik, 1978–95* (Düsseldorf: Kunstmuseum Düsseldorf, 1996).
12. Henri Bergson, *Time and Free Will: An Essay on the Immediate Data of Consciousness*, trans. F. L. Pogson (New York: Dover, 2001), 155.
13. "The will to power is not a being, not a becoming, but a *pathos*—the most elemental fact from which becoming and effecting first emerge." Nietzsche, *The Will to Power*, 339; my emphasis.
14. Bergson, *Time and Free Will*, 153.
15. Gabriel Tarde, *Essais et mélanges sociologiques* (Lyon: Storck, 1895), 237–38.
16. Henri Bergson, *Duration and Simultaneity*, trans. Mark Lewis and Robin Durie (Manchester: Clinamen, 1999), 30.
17. Bergson, *Time and Free Will*, 151–52. The actual-virtual relation that defines consciousness as memory not only produces this inorganic energy but produces and "utilizes in its own way" all the forms of energy known to physics. This force, by accumulating duration, thereby escapes the law of the conservation of energy and defines nonconservative states far from equilibrium. "But, if molecular movement can create sensation out of a zero of consciousness, why should not consciousness in its turn create movement either out of a zero of kinetic and potential energy, or by making use of this energy in its own way?" Bergson, 152.
18. "Subjectivity is set up at the intersection of sign fluxes and machinic fluxes, at the crossroads of facts of meaning, material and social facts, and, above all, of their transformations, resulting from their different modalities of assemblage." Félix Guattari, "The Schizoanalyses," in *Soft Subversions: Texts and Interviews, 1977–1985*, trans. Chet Wiener and Emily Wittman (Los Angeles: Semiotext(e), 2009), 209.
19. Walter Benjamin, "Paralipomênes et variantes de la seconde version," in *Ecrits français*, ed. Jean-Maurice Monnoyer (Paris: Gallimard, 1991), 184.
20. Walter Benjamin, "The Work of Art in the Age of Its Technological Reproducibility (Second Version)," in *Selected Writings*, vol. 3, *1935–1938*, ed. Howard Eiland and Michael W. Jennings (Cambridge: Harvard University Press, 2002), 118.
21. Benjamin, 117, 113.
22. Benjamin, 116, 119.

1. THE WAR MACHINE OF THE KINO-EYE

1. Dziga Vertov, "The Birth of Kino-Eye," in *Kino-Eye: The Writings of Dziga Vertov*, ed. Annette Michelson, trans. Kevin O'Brien (Berkeley: University of California Press, 1984), 41. Translation modified. All of the Vertov texts cited in this chapter come from this source.
2. Vertov, "Kinoks: A Revolution," 17–18.
3. Vertov, "Kinopravda," 131.
4. Vertov, "On *Kinopravda*," 45. Translation modified.

5. Vertov, "*Kinoglaz* (a Newsreel in Six Parts)," 39. Translation modified.
6. Vertov, "Kino-Eye," 72.
7. Vertov, "From the History of the Kinoks," 100.
8. Vertov, "*Kinoglaz* (a Newsreel in Six Parts)," 39.
9. Vertov, "Kino-Eye," 62. Translation modified.
10. Vertov, "Kino-Eye," 61. Translation modified.
11. Vertov, "Artistic Drama and Kino-Eye," 48.
12. Vertov, "From Kino-Eye to Radio-Eye," 88.
13. Vertov, "From Kino-Eye to Radio-Eye," 88.
14. Vertov, "From Kino-Eye to Radio-Eye," 91.
15. Vertov, "Artistic Drama and Kino-Eye," 47. Translation modified.
16. Vertov, "Kino-Eye," 75.
17. Vertov, "*Kinoglaz* (a Newsreel in Six Parts)," 39.
18. Vertov, "The Factory of Facts," 58.
19. Vertov, "From Kino-Eye to Radio-Eye," 88. Translation modified.
20. Vertov, "From Kino-Eye to Radio-Eye," 90.
21. Vertov, "WE: Variant of a Manifesto," 8.
22. Vertov, "WE: Variant of a Manifesto," 8. Translation modified.
23. Vertov, "Kinoks: A Revolution," 19.
24. Guy Debord, *Society of the Spectacle*, trans. Donald Nicholson-Smith (New York: Zone, 2004), 12, 24.
25. Vertov, "Without Words," 118–19. Translation modified.
26. Vertov, "Without Words," 118. Translation modified.
27. Vertov, "Kino-Eye," 75.
28. Vertov, "Kino-Eye," 71.
29. Vertov, "On the Film Known as *Kinoglaz*," 34.
30. Vertov, "Kino-Eye," 70–71. Translation modified.
31. Vertov, "*Three Songs of Lenin* and Kino-Eye," 125.
32. Vertov, "The Essence of Kino-Eye," 49–50.
33. Vertov, "Kino-Eye," 74.
34. Vertov, "From Kino-Eye to Radio-Eye," 92.

2. BERGSON AND MACHINES THAT CRYSTALLIZE TIME

1. According to Kazimir Malevich, the introduction of images and movement into the world is expressed in modern art by a "pulverization" of subject and object, and by the emergence of the energy forces that constitute them. Malevich's artistic trajectory overlaps, in a certain way, with Bergson's philosophical trajectory.

Futurism has elucidated the situation of one who represents the world that has been set into movement. The human forms a center around which this movement takes place. He establishes that such a phenomenon is not produced exclusively within a single, wedge-shaped radius of convergence, but in front and behind, on the sides, above, and below. The human is an axis around which a million mechanisms move and seeing all this does not mean seeing it with eyes, since it can be seen with knowledge and one's entire being. And since Futurism is only interested in translating into the omnimovement of force the organisms fluttering around the city, moving towards a global dynamism, capturing the general state of rotation, our psyche attaches to itself a new and real representation of the contemporary state of our understanding of the world, or better, within our psychic monad our brain reflects, as in a mirror, its real states. (Kazimir Malevitch, *De Cézanne au suprématisme* [Lausanne: L'âge d'homme, 1974], 107–9)

The movement of the city is simultaneously a movement of images: movement-image. "The petit bourgeois is struggling to navigate this global movement of the city, in which, furthermore, images are useful only insofar as their form is indispensable as mutual action, in contrast to the effort of dynamic expression" (Malevitch, 107–9). But the dynamism involved in "pulverizing" objects equally overthrows the movement-image, and the work of the artist must therefore aim at the "force of excitation" that grounds it: "The whole direction of art, expressing itself in movement, was also launched along a contemporary line of the development of this movement. And when, on this line, it attained a new and stronger tension, the forms themselves, as signs in movement, became other. It is clear that futurism, as a more powerful revelation, provided another image of movement, the assembled force of tension. And if in a subsequent movement the signs of energy develop further, the dynamometers of art will continue to increase" (ibid.). With Malevich, the dynamometer will even be able to grasp the "pure action" of sensation, excitation, and energy forces, the non-figurative conditions of representation.

2. Henri Bergson, "The Possible and the Real," in *The Creative Mind*, trans. Mabelle Andison (New York: Dover, 2010), 75.

3. "For want of a better word we have called it consciousness. But we do not mean the narrowed consciousness that functions in each of us. Our own consciousness is the consciousness of a certain living being, placed in a certain point of space.... In order that our consciousness shall coincide with something of its principle, it must detach itself from the *already-made* and attach itself to the *being-made*." Henri Bergson, *Creative Evolution*, trans. Arthur Mitchell (Basingstoke: Palgrave, 2007), 152–53. Individual consciousness, in its continuous streaming, introduces us to the interior of a reality on the model of which we must represent others to ourselves. It is only through individual consciousness that we can perceive something of ontological consciousness. But in no case can consciousness be reduced to an anthropomorphism. "Throughout the whole extent of the animal kingdom, we have said, consciousness seems proportionate to the living being's power of choice" (Bergson, 115). In fact,

Bergson speaks of a "supra-consciousness"—distinct from individual consciousness—of life, élan, the virtual, and pure will. All of these terms refer, in different ways, to a becoming that is "a growth from within, the uninterrupted prolongation of the past into a present which is already blending into the future." Bergson, *The Creative Mind*, 20.
4. Henri Bergson, *Mind-Energy*, trans. H. Wildon Carr (Basingstoke: Palgrave, 2007), 143.
5. Bergson, *Creative Evolution*, 192.
6. Ibid.
7. Henri Bergson, *Matter and Memory*, trans. Nancy Margaret Paul and W. Scott Palmer (New York: Zone, 1991), 208.
8. Ibid., 35.
9. "To perceive all the influences from all the points of all bodies would be to descend to the condition of a material object." Bergson, *Matter and Memory*, 49.
10. The solidification that comes by cooling the prebiological soup of pure perception—that contains the identity of matter, light, image, and time—gives rise to matter as we know it. But even in this case matter is a duration, a duration that is an infinite repetition or a present that eternally begins again. Matter is a scansion of independent, infinitely repeated moments. Matter, says Bergson, citing Gottfried Leibniz, is instantaneous mind. "If matter does not remember the past, it is because it repeats the past unceasingly, because, subject to necessity, it unfolds a series of moments of which each is the equivalent of the preceding moment and may be deduced from it: thus its past is truly given in its present." Bergson, *Matter and Memory*, 222–23.
11. Bergson, "The Possible and the Real," 76.
12. Bergson, *Matter and Memory*, 19.
13. Gilles Deleuze, *Cinema 1: The Movement-Image*, trans. Hugh Tomlinson and Barbara Habberjam (Minneapolis: University of Minnesota Press, 1986), 87.
14. Bergson, *Matter and Memory*, 57.
15. Bergson, 56.
16. The body is simply a conductor (i.e., without duration) of the instantaneous. It is therefore necessary to correct, in the following way, what Bergson has written about the ideal situation of pure perception: "But already we may speak of the body as an ever advancing boundary between the future and the past, as a pointed end, which our past is continually driving forward into our future." Bergson, *Matter and Memory*, 78.
17. Bergson, 205.
18. Bergson, 101.
19. "In short, it is a work of divination, but not of abstract divination; it is an externalization of memories, of perceptions simply remembered and consequently unreal, which profit by the partial realization that they find here and

there in order to be realized integrally. Thus, in the waking state, the knowledge we seize of an object implies an analogous operation to that which is accomplished in dream. We perceive only of the thing a mere sketch.... It is this kind of hallucination, inserted and fitted into a real frame, which we provide for ourselves when we perceive things." Bergson, *Mind-Energy*, 95.
20. Bergson, *Matter and Memory*, 102.
21. Henri Bergson, "The Perception of Change," in *The Creative Mind*, 128.
22. Bergson, *Mind-Energy*, 132.
23. "For consciousness there is no present, if the present be a mathematical instant. An instant is the purely theoretical limit which separates the past from the future. It may, in the strict sense, be conceived, it is never perceived." Bergson, *Mind-Energy*, 5.
24. Bergson, *Mind-Energy*, 132.
25. Gilles Deleuze, *Difference and Repetition*, trans. Paul Patton (New York: Columbia University Press, 1994), 73.
26. Bergson, *Matter and Memory*, 204.
27. Bergson, 198.
28. Bergson, "The Possible and the Real," 74–75.
29. Deleuze, *Difference and Repetition*, 87.
30. Deleuze, 80.
31. Gilles Deleuze, *Cinema 2: The Time-Image*, trans. Hugh Tomlinson and Robert Galeta (Minneapolis: University of Minnesota Press, 1989), 82–83.
32. For Bergson, a living being is a center of action. It represents the introduction of a certain amount of contingency, a certain quantity of possible action, into the world. However, this center should not be understood as a thing, but rather as a continuity of outpourings.
33. "Visual perception is nothing else: the visible outlines of bodies are the design of our eventual action on them." Bergson, *Creative Evolution*, 62.
34. "It was to create with matter, which is necessity itself, an instrument of freedom, to make a machine which should triumph over mechanism, and to use the determinism of nature to pass through the meshes of the net which this very determinism had spread." Bergson, *Creative Evolution*, 169. "But this consciousness, which is a need of creation, is made manifest to itself only where creation is possible. It lies dormant when life is condemned to automatism: it wakens as soon as the possibility of a choice is restored" (Bergson, 167).
35. "Consciousness is the light that plays around the zone of possible actions or potential activity which surrounds the action really performed by the living being. It signifies hesitation or choice." Bergson, *Creative Evolution*, 93.
36. In this context, electronic technology is a machine produced by human understanding that brings us a little bit closer to the real continuum of matter and allows us to penetrate a little bit more intimately into the connection that perception has with the fabrication of images through time.

37. Bergson, *Creative Evolution*, 102–3.
38. "Certainly, language would not have given the faculty of reflecting to an intelligence entirely externalized and incapable of turning homeward. An intelligence which reflects is one that originally had a surplus of energy to spend, over and above practically useful efforts. It is a consciousness that has virtually reconquered itself. But still the virtual has to become actual." Bergson, *Creative Evolution*, 102. Unfortunately, the understanding of this dual function of language is not shared by linguists or researchers who work with digital technologies: "The mental machine projects into consciousness, through images and words, what was or what could be. Digital simulation certainly uses language, but an already-formalized language filtered through the logic of calculation." Edmond Couchot, "L'Odyssée mille fois, ou Les machines à langage," in *Traverses* 44–5 (1988): 91. To make images and words (of language) the conditions of memory and imagination is a way to neutralize concrete time—"that which makes such that everything is made"—as well as a way to have a truncated vision of the process of creation (of images). This is why I prefer the phrase "machines that crystallize time" over the much more ambiguous "language machines" to define digital technologies.
39. Bergson, *Creative Evolution*, 103.
40. Bergson.
41. Gilbert Simondon, *L'individuation psychique et collective* (Paris: Aubier, 1989), 271.
42. Bergson, *Creative Evolution*, 114. But it must be emphasized that the conditions that make it possible to go beyond the opposition between intellect and life were created by the intellect itself: "Without intelligence, it would have in the form of instinct, riveted to the special object of its practical interest, and turned outward by it into movements of locomotion." Bergson, *Creative Evolution*, 115.
43. In fact, science has hardly integrated the Bergsonian critique of "abstract time" into its speculations and experiments. "The systems science works with are, in fact, in an instantaneous present that is always being renewed: such systems are never in that real, concrete duration in which the past remains bound up with the present." Bergson, *Creative Evolution*, 14. For a critique along these lines, see the work of Ilya Prigogine and Isabelle Stengers. In the social sciences the critique of structuralism begins—albeit with many contradictions—with the introduction of the event.
44. "But, supposing this view were finally to prevail, it should only lead, on deeper study, to some other mode of analyzing of the living being, and so to a new discontinuity—although less removed, perhaps, from the real continuity of life." Bergson, *Creative Evolution*, 105.
45. "Fabricating consists in shaping matter, in making it supple and in bending it, in converting it into an instrument in order to become master of it." Bergson, *Creative Evolution*, 118.

46. The apprehension of this new dimension requires us to go beyond and redefine the opposition between matter and mind: "We have repudiated materialism, which derives [inextension] from [extension]; but neither do we accept idealism, which holds that [extension] is constructed by [inextension]. We maintain, as against materialism, that perception overflows infinitely the cerebral state; but we have endeavored to establish, as against idealism, that matter goes in every direction beyond our representation of it." Bergson, *Matter and Memory*, 181. The limits of materialism and idealism "duplicate" material and moral states, according to the priorities given to matter or spirit. Bergson, however, situates himself at the confluence of these two degrees of reality, precisely where we pass from one to the other.
47. Bergson, *Matter and Memory*, 181.
48. In Bergson we also find another distinction between a "vast body" and a "small body":

> Yet, even physically, man is far from merely occupying the tiny space allotted to him. . . . For if our body is the matter to which our consciousness applies itself it is coextensive with our consciousness, it comprises all we perceive, it reaches to the stars. But this vast body is changing continually, sometimes radically, at the slightest shifting of one part of itself which is at its center and occupies a small fraction of space. This inner and central body, relatively invariable, is ever present . . . it is operative: it is through this body, and through it alone, that we can move other parts of the large body. And, since action is what matters, since it is an understood thing that we are present where we act, the habit has grown of limiting consciousness to the small body ignoring the vast one. . . . If the surface of our organized small body (organized precisely with a view to immediate action) is the seat of all our actual movements, our huge inorganic body is the seat of our potential or theoretically possible actions. (Henri Bergson, *Two Sources of Morality and Religion*, trans. R. Ashley Audra and Cloudesley Brereton [Notre Dame: University of Notre Dame Press, 1977], 258)

3. VIDEO, FLOWS, AND REAL TIME

1. Let us take the example of a technical means of expression that corresponds to the general deterritorialisation of flows in capitalism—the flow of electricity, which can be considered as the realization of any flow as such. Marshall McLuhan said it in a different way: "The electric light is pure information. It is a medium without a message." Marshall McLuhan, *Understanding Media: The Extensions of Man* (Cambridge: MIT Press, 1994), 8. All images and sounds produced by electronic and computer technologies are transformations and combinations (compositions) of the intensities, forces, and fields that move into the flow: electromagnetic flows for video, optical flows for

telematics, algorithmic flows for computing. The passage from the first to the last is marked by an increasing deterritorialization. Fiber optics replaces copper. With the advent of lasers and silica fiber, light could be channeled and controlled, thereby replacing electrical impulses as vectors of information in a network. Computer flows, in turn, free themselves from both matter and light by simply being a mathematical, nondiscursive language. But in any case, what is particularly interesting is the relationship between signifying and asignifying flows, which the new paradigm introduces. Images and sounds are produced by machines whose materials consist of a new matter that is in continuous modulation and perpetual variation and that constitutes these flows. Indeed, "these figures do not derive from a signifier nor are they even signs as minimal elements of the signifier; they are nonsigns, or rather nonsignifying signs, points-signs having several dimensions, flows-breaks or schizzes that form images through their coming together in a whole, but that do not maintain any identity when they pass from one whole to another. Hence the figures, that is, the schizzes or breaks-flows are in no way 'figurative'; they become figurative only in a particular constellation that dissolves in order to be replaced by another one." Gilles Deleuze and Félix Guattari, *Anti-Oedipus*, trans. Robert Hurley, Mark Seem, and Helen Lane (Minneapolis: University of Minnesota Press, 1983), 241. These are semiotics, languages, and codes that no longer pass through human subjectivity but open themselves up to other processes of subjectivation and to other becomings. The language of informatics does not pass through writing or the voice but operates without them both. "The computer is a machine for instantaneous and generalized decoding" that functions on and through digital "flows-breaks or schizzes" (Deleuze and Félix Guattari, 241).

2. It is noteworthy that Bergson, limiting himself to a critique of the illusion of movement, did not grasp the force of the crystallization of time that was already active in cinema: "For the first time in the history of the arts, in the history of culture, man found the means *to take an impression of time*. And simultaneously the possibility of reproducing that time on screen as often as he wanted, to repeat it and go back to it. He acquired a matrix of *real time*. Once seen and recorded, time could now be preserved in metal boxes over a long period (theoretically forever)." Andrei Tarkovski, *Sculpting in Time: Reflections on the Cinema*, trans. Kitty Hunter-Blair (Austin: University of Texas Press), 62. But it is only with video that this "matrix of real time" finds an adequate technological assemblage.

3. Angela Melitopoulos, "Vidéo, temps et mémoire," interview by Maurizio Lazzarato, *Chimeras* 27 (Winter 1996): 95.

4. Despite the claim made by Bergson, who compared human perception to the illusion of movement created by the cinematograph, in montage movement is not given to images through the projection apparatus. The cinematograph,

according to Bergson, takes snapshots of reality (the photogram) and then reconstructs motion by adding a mechanical movement to the images. What Bergson did not see is that montage is the element that distributes a past and a future in film, the real emergence of time.

5.

In video, unlike cinema, it is not the succession of shots that creates movement, but the movement of light itself. . . . I do not work with the succession of shots, but with light. . . . If I fail, it is because I have lapsed into cinematographic schemes. When I started working with video, I noticed that movement is not about simply moving something in space. . . . I only discovered this with video, before that I did not think about it. . . . By slowing down the video image, by dilating and stretching it, I saw that movement did not decrease but that, on the contrary, it might accelerate since the grains moved differently. In cinema, slow motion is a slower succession of shots, but this has nothing to do with video. (Melitopoulos, "Vidéo, temps et mémoire," 96)

6. "An Interview with Bill Viola," by Raymond Bellour, *October* 34 (Autumn 1985): 100.
7. "An Interview," 100.
8. "An Interview," 100.
9. Henri Bergson, *Matter and Memory*, trans. Nancy Margaret Paul and W. Scott Palmer (New York: Zone, 1991), 165.
10. Gilles Deleuze, *The Fold*, trans. Tom Conley (Minneapolis: University of Minnesota Press, 1993), 77.
11. Bergson, *Matter and Memory*, 70.
12. Bergson, 203.
13. Nam June Paik, *Du cheval à Christo et autres écrits* (Paris: Lebeer Hossman, 1993), 110.
14. Bill Viola, "The Sound of One Line Scanning," in *Reasons for Knocking at an Empty House*, ed. Robert Violette (Cambridge: MIT Press, 1995), 158.
15. The labor of the body and brain is automatic; memory has no freedom, since it is completely occupied by the realization of an action. Memory contracts and expands time, but it is a habit that is repeated. Memory cannot begin to produce images freely and becomes intellectual labor only when it has succeeded in extricating itself from the necessity of the finalized action. It can then introduce indeterminacy, the unpredictable, into the process of perception. In the same way, the video recorder and more sophisticated techniques of image processing enable greater freedom in the contraction-relaxation of flows.
16. Nam June Paik, "Input-Time and Output-Time," in *Video Art: An Anthology*, ed. Beryl Korot and Ira Schneider (New York: Harcourt Brace, 1977), 98. More subjectively, Angela Melitopoulos explains:

> Through my work with video, I construct something that is perhaps a memory, a second level of memory that is connected to the machine, especially since I understand my own memory only through video. I can reconstruct a chain of memories around a lived, but forgotten, experience. I begin to shoot and establish relationships with my lived experience. Then chains of images arise and I can move forward or backward. I use different regimes of parallel images within which I evolve. As time passes, other things come up and I can use them to construct parallel paths, similar memories that sometimes seem more accurate than "real" memories and, with their vision, my memory is solicited differently. It is through this recording, and the work it involves, that I can create my own memory or, at least, something in my memory. (Melitopoulos, "Vidéo, temps et mémoire," 96)

17.

> We must not say montage but process, video-processing (as in English), when defining the work on the electronic structure and the image. Montage belongs to cinematic language. It means putting together the shots and sounds. Processing is working on the image itself. This means that the image begins somewhere, changes, becomes something else, and something else. It is not at the beginning that you should see the image, it is not pre-defined. The image has different possible transformations. An image, you can work on it for years, to infinity. . . . You can extend it, slow it down, accelerate it, take a piece and forget the rest. That is what is different from cinema. To take everything together, as in cinema, does not interest me. (Melitopoulos, "Vidéo, temps et mémoire," 97)

18. Bill Viola, *Video* (Paris: Seuil/Communications, 1988), 72.
19. Viola, 74.
20. Paik, *Du cheval à Christo et autres écrits*, 145.
21. Sergei Eisenstein, *Notes of a Film Director*, trans. X. Danko (Moscow: Foreign Languages, 1958), 6.
22. Walter Benjamin, "Paralipomênes et variantes," in *Ecrits français*, ed. Jean-Maurice Monnoyer (Paris: Gallimard, 1991), 182.
23. Bergson, *Matter and Memory*, 205, 208.
24. The Bergsonian conceptualization can still be useful here. According to Bergson, perception, however rapid it is supposed to be (as fast, for example, as the speed of light in real time), always occupies a certain thickness of duration.
25. Paik, *Du cheval à Christo et autres écrits*, 87.
26. Paik, 150.
27. Paik, 132. It is interesting to note that around the same time situationist theory was developing, which in the beginning consisted of the practice of situations.
28. Antonio Negri, *Macchina tempo* (Milan: Feltrinelli, 1980).

29. I am obviously talking about the production of collective images, which will increasingly depend on machines that crystallize time.
30. Television produces time signs that are inseparable from word signs, thought signs, and affect signs. But this is a domain of the operation of television that cannot be analyzed here.
31. The specificity of its form of representation consists in its material: the images, durations, times.
32. Wolf Vostell, "Entretien avec Wolf Vostell," *Cahiers du Cinema* 332 (February 1982).

4. BERGSON AND SYNTHETIC IMAGES

1. Edmond Couchot, "L'Odyssée mille fois ou les machines à langage," *Traverses* 44–45 (August 1988): 90–91.
2. Couchot, 90.
3. "The truth is, when one speaks of the gradual formation of the eye, and, still more, when one takes into account all that is inseparably connected with it, one brings in something entirely different from the direct action of light. One implicitly attributes to organized matter a certain capacity *sui generis*; the mysterious power of building up very complicated machines to utilize *the simple excitation* that it undergoes." Henri Bergson, *Creative Evolution*, trans. Arthur Mitchell (Basingstoke: Palgrave, 2007), 46 (my emphasis). According to Bergson, the role that light plays in perception must be interpreted differently. In order to "utilize the simple excitation" that is light, life has no other solution than to adapt to it, at first passively: "Where it has to direct a movement, it begins by adopting it. Life proceeds by insinuation" (46). But from this genealogy we should not draw the conclusion that the action of light and the eye are of the same nature, distinguished only by a level of complexity. Between this initial movement and the eye there is the same interval as between a photograph and a camera: "Certainly the photograph has been gradually turned into a photographic apparatus; but could light alone, a physical force, ever have provoked this change, and converted an impression left by it into a machine capable of using it?" (46). The theses that define the transition from analog to digital in relation to the role that light plays in the production of the image take part in the misunderstanding of the relationship between the "simple excitation" operated by light and the "sui generis" force that arises from highly complex machines (from the eye to electronic and digital technologies).
4. Gabriel Tarde, *Essais et mélanges sociologiques* (Lyon: Stock, 1895), 237.
5. Henri Bergson, *Mind-Energy*, trans. H. Wildon Carr (Basingstoke: Palgrave, 2007), 149.
6. It is curious to note that D. H. Lawrence, speaking about a form of knowledge (immediate, profound, direct, by instinct or intuition) that we have lost, refers to the game of chess: "It was a knowledge based not on words but on images. The abstraction was not into generalizations or into qualities, but

into symbols. And the connection was not logical but emotional. . . . Images or symbols succeeded one another in a procession of instinctive and arbitrary physical connection . . . and they 'get nowhere' because there was nowhere to get to, the desire was to achieve a consummation of a certain state of consciousness. . . . Perhaps all that remains to us today of the ancient way of 'thought-process' are games like chess and cards." D. H. Lawrence, *Apocalypse* (London: Penguin, 1996), 91.

7. On the concept of unity and multiplicity, and their relationships: "I am then (we must adopt the language of the understanding, since only the understanding has a language) a unity that is multiple and a multiplicity that is one; but unity and multiplicity are only views of my personality taken by an understanding that directs its categories at me; I enter neither into one nor into the other nor into both at once, although both, united, may give a fair imitation of the mutual interpenetration and continuity that I find at the base of my own self. Such is my inner life, and such also is life in general." Bergson, *Creative Evolution*, 165.

8. Edmond Couchot, among others, maintains this misunderstanding: "Neither real nor imaginary, the world of digital simulation belongs to another category: *the virtual.*" Couchot, "L'Odyssée mille fois ou les machines à langage," 92.

9. "While, in its contact with matter, life is comparable to an impulsion or an impetus, regarded in itself it is an immensity of potentiality, a mutual encroachment of thousands and thousands of tendencies which nevertheless are 'thousands and thousands' only when once regarded as outside of each other, that is, when spatialized." Bergson, *Creative Evolution*, 165–66.

10. Actualization is a differentiation of time, a split into two symmetrical streams, the possible and the real, one of which falls back toward the past while the other leaps toward the future.

11. Bergson, *Creative Evolution*, 120.

12. Henri Bergson, "Introduction to Metaphysics" in *The Creative Mind*, trans. Mabelle Andison (New York: Dover, 2010), 158–59.

13. The transition from analog to digital is interpreted here as a transition from perception to imagination.

14. Bergson, *Creative Evolution*, 129.

15. "An Interview with Bill Viola," by Raymond Bellour, *October* 34 (Autumn 1985): 114.

16. "An Interview," 116.

17. Edmond Couchot, *De l'optique au numérique* (Paris: Hermes, 1988), 210–11.

18. Couchot, 210.

19. Jean-Louis Weissberg, "Sous les vagues, la plage," in *Paysages virtuels: Image vidéo, image de synthèse* (Paris: Dis voir, 1988), 47–48.

20. Claude Cadoz, *Les réalités virtuelles* (Paris: Flammarion, 1994), 85.

5. NIETZSCHE AND TECHNOLOGIES OF SIMULATION

1. Friedrich Nietzsche, *La volonté de puissance* (Paris: Librairie générale française, 1991), 325.
2. Nietzsche, 71.
3. See chapter 6.
4. Friedrich Nietzsche, *The Gay Science*, trans. Josefine Nauckhoff (Cambridge: Cambridge University Press, 2001), 63.
5. Friedrich Nietzsche, *On the Genealogy of Morality*, trans. Carol Diethe (Cambridge: Cambridge University Press, 2006), 87.
6. Nietzsche, *La volonté de puissance*, 293.
7. Nietzsche, 241.
8. "This phenomenon is already recognizable in vision. All the senses freely produce their impressions; most of sensory perception is a divination." Nietzsche, *La volonté de puissance*, 241.
9. Bergson also recognizes the importance of aesthetic forces in vision:

   That an effort of this kind is not impossible is proved by the existence in man of an aesthetic faculty along with normal perception. Our eye perceives the features of the living being, merely as assembled, not as mutually organized. The intention of life, the simple movement that runs through the lines, that binds them together and gives them significance, escapes it. This intention is just what the artist tries to regain, in placing himself back within the object by a kind of sympathy, in breaking down, by an effort of intuition, the barrier that space puts up between him and his model. It is true that this aesthetic intuition, like external perception, only attains the individual. (Henri Bergson, *Creative Evolution*, trans. Arthur Mitchell [Basingstoke: Palgrave, 2007], 114)

10. Friedrich Nietzsche, *Le livre du philosophe* (Paris: Flammarion, 1991), 58.
11. Nietzsche, 93.
12. Nietzsche, *La volonté de puissance*, 250.
13. Nietzsche, 338.
14. Nietzsche, 314.
15. Friedrich Nietzsche, *Beyond Good and Evil*, trans. Judith Norman (Cambridge: Cambridge University Press, 2002), 82.
16. Nietzsche, *La volonté de puissance*, 310.
17. Real time seems to be, first of all, a translation by digital technologies of the abstract time with which science denies concrete time (true real time).

    What does it mean, to say that the state of an artificial system depends on what it was at the moment immediately before? There is no instant immediately before another instant: there could not be any more than there could be one mathematical point touching another. The instant "immediately before" is, in reality, that which is connected with the present instant by the interval dt. All that you mean to say, therefore, is that the present state of the system is defined by equations into which differential coefficients enter, such as ds/dt, dv/dt, that is to say, at bottom, present velocities and present accelerations. You are therefore really speaking only of the present—a present, it is true, considered along with its tendency. The systems science works with are, in fact, in an instantaneous present that is always being renewed: such systems are never in that real concrete duration in which the past remains bound up with the present. (Bergson, *Creative Evolution*, 14)

Real time is therefore the realization of abstract time. But consider the case of the computer's clock. According to chapter 2, this mechanical clock, constructed on and by discrete elements, brings us closer to real duration, to concrete time. The frequency of the rhythm at which the quartz clock analyses a second is so high that it might be compared to a wave, even if this wave is composed of the tiniest possible elements. From this point of view, the definition of real time is correct because it allows us to approach more closely the continuity that constitutes concrete time. This possibility is denied not by technology (oriented, within the limits of its construction, in a direction that makes it possible to accumulate "pure vibrations") but by the social temporality organized by the abstract time of capitalism.

18. Nietzsche, *La volonté de puissance*, 290.
19. Nietzsche, *The Gay Science*, 113.
20. Nietzsche, *La volonté de puissance*, 288.
21. This definition, common in the early stages of research into artifical intelligence, is identical to the one Bergson uses to describe the reduction involved in the operation of the logical model and the limits this model places upon the apprehension of thought and action.

It must, therefore, in order to think itself clearly and distinctly, perceive itself under the form of discontinuity. Concepts, in fact, are outside each other, like objects in space; and they have the same stability as such objects, on which they have been modelled. Taken together, they constitute an "intelligible world," that resembles the world of solids in its essential characters, but whose elements are lighter, diaphanous, easier for the intellect to deal with than the image of concrete things: they are not, indeed, the perception itself of things, but the representation of the act by which the intellect is fixed on them. They are, therefore, not images, but symbols. Our logic is the complete set of rules that must be followed in using symbols. (Bergson, *Creative Evolution*, 103–4)

22. "Kasparov was beaten in his first game against Deep Blue, but did not ultimately lose. To think that humans operate by mere calculation presupposes very unlikely assumptions. Despite the tremendous calculating power of machines, he is still victorious. What is extraordinary is that, in these conditions, machines have not yet managed to beat humans. How do they operate, so that this might become the basis of computer programming?" Jacques Arsac, "Kasparov contre Deep Blue," *La Recherche* 289 (1996).
23. Nietzsche, *La volonté de puissance*, 41.
24. Nietzsche, *The Gay Science*, 112–13.
25. Nietzsche, 185–86.
26. Mony Ekaim and Isabelle Stengers, "Du mariage des hétérogènes," *Chimères* 21 (Winter 1994): 149.
27. Francisco Varela, "Pour une nouvelle sciences cognitive," *Chimères* 28.
28. Varela.
29. Varela.
30. Varela.
31. "Everything is obscure in the idea of creation if we think of things which are created and a thing which creates, as we habitually do, as the understanding cannot help doing. We shall show the origin of this illusion in our next chapter. It is natural to our intellect, whose function is essentially practical, made to present to us things and states rather than changes and acts. But things and states are only views, taken by our mind, of becoming. There are no things, there are only actions." Bergson, *Creative Evolution*, 159.
32. Nietzsche, *La volonté de puissance*, 355–56.
33. Nietzsche, 288–89. The term *simulation* is often used by Nietzsche when discussing all the fictions with which we think.

6. THE ECONOMY OF AFFECTIVE FORCES

1. Henri Bergson, *Time and Free Will*, trans. F. L. Pogson (New York: Dover, 2001), 152, 211.
2. Gilles Deleuze and Félix Guattari, *A Thousand Plateaus*, trans. Brian Massumi (Minneapolis: University of Minnesota Press, 1987), 492.
3. Gilles Deleuze and Félix Guattari, *Anti-Oedipus*, trans. Robert Hurley, Mark Seem, and Helen Lane (Minneapolis: University of Minnesota Press, 1983), 3–4.
4. Félix Guattari, *Schizoanalytic Cartographies*, trans. Andrew Goffey (London: Bloomsbury, 2013), 39.
5. Félix Guattari, *La révolution moléculaire* (Paris: Recherches, 1977), 80.
6. Bergson, *Time and Free Will*, 32.
7. Guattari, *La révolution moléculaire*, 93. Guattari confirms and deepens this analysis in a later work: "Just as social machines can be grouped under the general title of collective facilities, technological machines of information and

communication operate at the heart of human subjectivity, not only within its memory and intelligence, but within its sensibility, affects, and unconscious fantasms." He describes taking into account the machinic dimensions of the production of subjectivity as follows: "Thus one finds in it: 1. Signifying semiological components which appear in the family, education, the environment, religion, art, sport. 2. Elements constructed by the media industry, cinema, etc. 3. Asignifying semiological dimensions that trigger informational sign machines, and that function in parallel or independently of the fact that they produce and convey significations and denotations, and thus escape from strictly linguistic axiomatics." Félix Guattari, *Chaosmosis: An Ethico-aesthetic Paradigm*, trans. Paul Bains and Julian Pefanis (Bloomington: Indiana University Press, 1995), 4. Translation modified.
8. Félix Guattari, *Soft Subversions*, trans. Chet Wiener and Emily Wittman (Los Angeles: Semiotext(e), 2009), 244, 258.
9. Deleuze and Guattari, *A Thousand Plateaus*, 458.
10. Guattari, *La révolution moléculaire*, 237.
11. Guattari, 573.
12. "The viewer is no more than the on-off synapse of a communication satellite that sends and receives multiple networks composed of millions of incoming and outgoing signals, impalpable pulsations of information." Nam June Paik, *Du cheval à Christo et autres écrits* (Paris: Lebeer Hossman, 1993), 145.
13. Guattari, *Chaosmosis*, 16.
14. Guattari, 17.
15. Guattari, 19.
16. To measure the full extent of this modelization, it must be juxtaposed with reception theory in all its versions: the Marxist tradition of media analysis (Stuart Hall), "weak semiotics" (Umberto Eco), the philosophy of perception (Hans Robert Jauss), and the whole tradition of the sociology of media. The differences between these theories are less important than what they have in common. They share the same ontology: the certainty of the distinction between human and machine, between subject and object, between words and things, between sensible and intelligible. But they do not suspect for a moment that we have passed over to the other side of the mirror, that the plane upon which the problem of reception arises is radically other, irreducible to that of language and communication.
17. Bernard Cache, "Terre meuble," in *Imaginaire technique*, ed. Cyrille Simonnet (Marseille: Parenthèses, 1997), 116.
18. Robert Fogel, *I&T Magazine*, 1995.
19. Fogel.
20. The development of techonomics, along with its technological conditions, indicates a shift in the hegemony of economic innovation from Japan to the United States. The 1980s had given the illusion of the supremacy of the Toyotist model. In reality, this model was a rationalization and development of the

well-known and already-mastered Fordist model. "The Japanese miracle was based on an exquisite refinement of industrial production." If on the one hand, the innovations of the Toyotist model opened up a new mode of production, they were nevertheless rooted in the world of material production. The changes that took place after the first oil crisis and accelerated in the 1980s were more in the direction of the constitution of immaterial production, understood as the production of subjectivity. "Japan, on the other hand, which emerged only a short time ago as the champion of the electronic industry, has suddenly been overshadowed in the world scene due to the shift from a material to an immaterial society." G. Santucci, "L'économie immatérielle," *I&T Magazine*, 1994, 47.

21. *Business Week*, June 1994, 36. Unlike the service economy, in the information economy wages are on average higher, as are wage increases and productivity growth; investments are increasing considerably (investment in information technology already exceeds investment in mechanical equipment); and products are more easily exportable. The export of information technology and equipment far exceeds the overseas sales of the aeronautics industry, with the United States emerging as the world's largest exporter of software. The information economy certainly has a greater impact than what normal economic indicators can measure, since it determines a ghost economy that goes beyond political economy. "The role of information is transforming the nature of the economy," according to Kenneth J. Arrow. In this area, the United States "is showing the way to the rest of the world."
22. *Business Week*, March 1994, 31.
23. *Business Week*, December 1994, 26.
24. *Business Week*, 26.
25. *Business Week*, March 1995, 39.
26. *La Tribune des Fossées*, December 1994.
27. Edward McCraken, *Business Week*, March 1994, 45.
28. McCraken, 45.
29. Capitalism has introduced a rupture not only within forms of production but also, simultaneously, within assemblages of the production of subjectivity that, according to Guattari, must never be separated from them. Indeed, according to Guattari, archaic forms of enunciation were based upon speech and direct communication, while new assemblages increasingly resort to informative and mediated flows, carried by machinic channels that overflow, on all sides, the former individual and collective subjective territories. "Whilst territorialized enunciation was logocentric and implied a personal mastery of the ensembles that it discursivized, deterritorialized enunciation, which can be characterized as machinocentric, relies on non-human procedures and memories to deal with semiotic complexes that for the most part escape from direct conscious control." Félix Guattari, *Schizoanalytic Cartographies*, 20.
30. Telecom Italia, internal document, 1994.
31. G. Ginder, "Y-a-t-il une vie après la télévision?," unpublished manuscript.

32. Guy Debord, *Society of the Spectacle*, trans. Donald Nicholson-Smith (New York: Zone, 1995), 130.

33. Evaluations of the economic and social impact of this post-industrial economy, grouped together by the press under the name "Cool Britannia" (publishing, design, software, fashion, radio-TV, advertising, music, film, computer games, architecture), are very difficult to assess since the statistical systems are poorly adapted. According to estimates, however, "Cool Britannia" represents between £50 and £100 billion, or between 8% and 16% of GNP. Games and music production account for 25% and 18% of the world market, respectively. The United Kingdom ranks first, before the United States, in exports of television series and novels. Recording and fashion houses, grouped together, are bigger employers than the steel and automotive industries combined. And they report more revenue than North Sea oil and gas or agricultural products. (*Le Monde,* March 5, 1998)

34. Deleuze and Guattari, *A Thousand Plateaus*, 397.

7. THE CONCEPT OF COLLECTIVE PERCEPTION

1. Walter Benjamin, "The Work of Art in the Age of Its Technological Reproducibility (Second Version)," in *Selected Writings*, vol. 3, *1935–1938*, ed. Howard Eiland and Michael W. Jennings (Cambridge: Harvard University Press, 2002), 104.
2. It happens in exceptional cases that consciousness suddenly abandons its attention to life and thus breaks its subordination to practical action within the sensory-motor schema: "Immediately, as though by magic, the past once more becomes present. In people who see the threat of sudden death unexpectedly before them, in the mountain climber falling down a precipice, in drowning men, in men being hanged. . . . That is enough to call to mind a thousand different 'forgotten' details and to unroll the whole history of the person before him in a moving panorama." Henri Bergson, "The Perception of Change," in *The Creative Mind*, trans. Mabelle Andison (New York: Dover, 2010), 127.
3. Walter Benjamin, "On the Concept of History," in *Selected Writings*, vol. 4, *1938–1940*, ed. Howard Eiland and Michael W. Jennings (Cambridge: Harvard University Press, 2003), 391.
4. Benjamin, 396, 390, 396.
5. "The desire of the present-day masses to 'get closer' to things, and their equally passionate concern for overcoming each thing's uniqueness by assimilating is as a reproduction." Benjamin, "The Work of Art," 105.
6. Benjamin, 119.

7. Mikhail Bakhtin, "Epic and Novel," in *The Dialogic Imagination: Four Essays*, trans. Caryl Emerson and Michael Holquist (Austin: University of Texas Press, 1981), 20.
8. Bakhtin, 19. According to Bakhtin, high literature of the classical period is projected onto the past, which "does not mean, of course, that there is no movement within. On the contrary, the relative temporal categories within it are richly and subtly worked out ... there is evidence of a high level of artistic technique in matters of time. But within this time, completed and locked into a circle, all points are equidistant from the real, dynamic time of the present; insofar as this time is whole, it is not localized in an actual historical sequence; it is not relative to the present or to the future; it contains within itself, as it were, the entire fullness of time" (Bakhtin, 19).
9. Bakhtin, 23.
10. Bakhtin, 37.
11. Giorgio Agamben, "Pour une éthique du cinéma," *Trafic* 3 (1993): 51. According to Agamben, it is for these reasons that modern theater has felt the need to avoid commedia dell'arte actors, though "not without holding onto their innovations." And it is in cinema that the disturbing prophecy of their bodies was to be accomplished.
12. Walter Benjamin, "Little History of Photography," in *Selected Writings*, vol. 2, *1927–1934*, ed. Michael W. Jennings, (Cambridge: Harvard University Press, 1999), 510–12.
13. For Bergson, morbid or abnormal psychological phenomena "appear to add something to normal life and enrich it instead of impoverishing it. A delirium, a hallucination, an obsession are positive facts.... They seem to introduce into the mind certain new ways of feeling and thinking." Henri Bergson, *Mind-Energy*, trans. H. Wildon Carr (Basingstoke: Palgrave, 2007), 121.
14. Benjamin, "The Work of Art," 120.
15. Walter Benjamin, "The Work of Art in the Age of Its Technological Reproducibility (First Version)," trans. Michael W. Jennings, *Grey Room* 39 (Spring 2010): 34.
16. Walter Benjamin, "The Formula in Which the Dialectical Structure of Film Finds Expression," in Eiland and Jennings, *Selected Writings*, 3:94.
17. Walter Benjamin, "The Author as Producer," in Jennings, *Selected Writings*, 2:771.
18. Benjamin, 771–72.
19. Walter Benjamin, "Paralipomênes et variantes," in *Ecrits français*, ed. Jean-Maurice Monnoyer (Paris: Gallimard, 1991), 176.
20. Benjamin, "Little History of Photography," 523.
21. Walter Benjamin, "La contribution d'Adorno à la discossion sur le fond," in Monnoyer, *Ecrits français*, 136.
22. Benjamin, 158.

23. Benjamin, "Paralipomènes et variantes," 181.
24. Benjamin, 169.
25. Félix Guattari shows, through the example of the Concorde (with a critical nod to the Heideggerian conception of technology), the plurality of elements involved in the realization of a technological apparatus, particularly the importance of economic and political elements. "The ontological consistency of this object is essentially composite; it is at the intersection, at the point of constellation and pathic agglomeration of Universes each of which have their own ontological consistency, traits of intensity, their ordinates and coordinates, their specific machinisms.... But the bottom line is that the ensemble of these final, material, formal, and efficient causes will not do the job! The Concorde object moves effectively between Paris and New York but remains nailed to the economic ground. This lack of consistency of one of its components has decisively fragilized its global ontological consistency." Félix Guattari, *Chaosmosis: An Ethico-aesthetic Paradigm*, trans. Paul Bains and Julian Pefanis (Bloomington: Indiana University Press, 1995), 47–48.
26. It is never a question of substitution, but always of a domination that incorporates other apparatuses, other practices, by creating other virtualities.
27. Gilles Deleuze defines the image of the cinema in the interwar period as "movement-image," "action-image."
28. But the cinema is not only that: "For the majority of city dwellers, throughout the workday in offices and factories, have to relinquish their humanity in the face of an apparatus. In the evening these same masses fill the cinemas, to witness the film actor taking revenge on their behalf not only by asserting his humanity (or what appears to them as such) against the apparatus, but by placing that apparatus in the service of his triumph," Benjamin, "The Work of Art," in *Selected Writings*, 3:111.
29. It is evident that this "free" temporality simply defines a new area of contestation.
30. Walter Benjamin, "Experience and Poverty," in Eiland and Jennings, *Selected Writings*, 2:732.
31. Walter Benjamin, "The Destructive Character," in Eiland and Jennings, *Selected Writings*, 2:542.

AFTERWORD: VIDEOPHILOSOPHY NOW

1. Many thanks to Angela Melitopoulos, who was present and helped mediate the discussion.
2. For an analysis of *Assemblages*, see Jay Hetrick, "Video Assemblages: 'Machinic Animism' and 'Asignifying Semiotics' in the Work of Melitopoulos and Lazzarato," *Footprint* 14 (Spring 2014): 53–68.
3. Introduction to this volume.

INDEX

Absolute past, 206
Abstract time, 60–61, 260n17
Acceleration, 96–97
Accumulation, of capital, 7
Accumulation-conservation, of time, 50
Action, 42–43; of artificial intelligence, 162–63; free, 196–97; knowledge and, 159–62; perception and, 69; political, xxiv; teleological, 71–72, 75; thinking and, 162–63; thought and, 152, 154–55
Action-reaction, 43–44, 48
Active force, 145–46
Activity, 41–42, 121
Actual, virtual and, 56–57, 60, 71–72, 105–6, 126, 129, 244n17
Actual image, 59–60, 67–68
Actual-virtual circuit, 64–65, 68, 102, 104–6, 111, 114, 127, 135, 225
Adorno, Theodor, 209, 214–15
Affect, 64–65, 172–74
Affection, 48–49, 64–65
Affective energy, 70, 73–74
Affective force, 49; of attention, 119–20; of machines that crystallize time, 169–70; production of images, 117–29
Agamben, Giorgio, 208–9, 267n11

Alterity, 151
Analog, 113–16
Ancient Greeks, 2
Animal, 78
Animism, 234
Apparatuses: of art, 226; of capture, 185; digital, 222–23; of power, 94, 101
Appearance, 143–44
Apperception, 211
Art: apparatuses of, 226; commercialization and, 209–10; force of, 147; masses absorbing, 205–6; political action and, xxiv; technological reproduction of, 200. *See also* Video art
Art and Multitude (Negri), xxiv–xxv
Artificial intelligence, 152, 154, 158–59, 161–63, 261n21
Artist, 150, 235
Artistic faculty, 147
Artistic practices, as technologies of self, xxiv
Artistic production, 193–94
Asignifying flows, 37, 180
Asignifying semiotics, xxii–xxiv, 176
Assemblage, xxii, 100, 160, 181, 192
Assemblages (2010), xxii, 228
Attention, 70–71, 119–20

Attention economy, 184, 189
Attentive perception, 52–53
Attentive recognition, 50–52
Auschwitz, 217, 219
Authorship, 31
Automatic recognition, 50–52, 55
Automation, 78, 152–53, 212
Automatism, 153, 158
Automaton, 154, 158

Bachelard, Gaston, 240n11
Bakhtin, Mikhail, 206–9, 266n8
Baudelaire, Charles, 200–1
Becoming, 101
Becoming-active, 213
Becoming-artist, 150
Being: appearance and, 143–44; image, as fabric of, 5; nothingness and, 4–5
Belief, 143
Benjamin, Walter, x, 206–9; on acceleration, 96; Adorno on, 214–15; on Bergson, 200–1, 203; on cinema, xv–xvi, 210–14; on collective perception, 199–200, 204–5, 215–16; on labor, perception, 1–2; on loss of experience, 177; on manual, intellectual labor, 17, 222; on new barbarism, 202–3, 226, 232–33; on power-time, 226; on revolution, 216; on technological reproduction, of work of art, 200; on virtual memory, 202; on worker, 221
Bergson, Henri, x–xv, xix, 242n2, 243n4; on affection, 48–49; on apprehension, of perception, 52–53; on attention, 70–71, 119; on attentive, automatic recognition, 50–52; Benjamin on, 200–1, 203; on brain, 46–47, 66, 106–7; on capacity for seeing, 11; on cinema, 82–83; on cinematograph, 253n4; on concept, percept, 2; on conceptualization, 113–14; on consciousness, 247n3, 249nn34–35, 266n2; on construction, of image, 117; on creation, 75–77; on digital technologies, 112; on duration, 14–15; duration time and, 37–39, 41; on force, 13–15; on force, time and, 40–41, 61–62; on image, of pure perception, 85; on image, time, 3–4; on intellectual labor, 240n17; on language, 72; Marx and, 229; *Matter and Memory* by, xv, 38, 120–21, 228; mechanical mysticism of, 77; on memory, 41–42, 50–53, 176, 201; on movement, 63; Nietzsche and, x, 13–14, 16, 229–30; on object, subject, 44–45; ontology of, 10, 182; on optical model, of production of images, 54–55, 211; on organism, 78–79; on perception, 2–3, 63, 87–88, 146, 148; on pure memory, 56–65; on pure perception, 44; on real versus abstract time, 60–61; on sensation, over perception, 48; on social memory, 102; Spinoza and, 231–32; on succession, of time, 59; on temporalities, 87–88; temporal ontology of, 1; temporal theory, on production of images, 53–54; on time, 12–13; *Two Sources of Morality and Religion* by, xiv, 242n2; on unforeseeable novelties, 8–9; on vast versus small bodies, 251n48; on virtual image, 130; on vision, 259n9; on will, 2
Beuys, Joseph, 100
Biopower, 173
Bloch, Ernst, xiv–xv
Body: brain and, 46; camera and, 21; collective, 35; as conductor, 248n16; as coordination, hierarchization of living beings, 164–67; as duration, 50; image and, 11, 45–46, 48; living,

79; memory and, 50–53; movement and, 125; as multiplicity, 164–67; Nietzsche on, 5–7, 164–67; pure perception and, 46; reality and, 6–7; sensation for, 48–49; sensory-motor activity of, 51–52, 91; in society, 165–66; spirit and, 6–7; syntheses of, 164–66; temporality of, 49; thoughts and, 164; vast versus small, 251n48
Brain: Bergson on, 46–47, 66, 106–7; body and, 46; consciousness and, 153–54; eye and, 55; image and, 50–51; movement and, 46–47, 50–51; video and, 66, 88

Cache, Bernard, 182
Cage, John, 101
Camera: body and, 21; cinema, 83; image and, 82; as input-output unit, 99; Life Caught Unawares by, 26–27; as machinic eye, 25; movement and, 28; Vertov on, 81. *See also* Video camera
Capacities, to act and feel, 40–41
Capital: accumulation of, 7; constant, 178, 198; ontology of, 10; postmodern, xxiv; striated, smooth, 197; valorization of, 170
Capitalism: affects, desires in, 172–74; asignifying semiotics in, xxii–xxiv, 176; creation in, 76; Deleuze and Guattari on, 170; deterritorialization by, 102, 174, 177; divisions in, 172; exploiting labor, society, 173; flows of knowledge, signs in, 174; Guattari on, xxii, 170, 173–75; as integral of power formations, 176; machines that crystallize time in, 175, 203; machinic enslavement in, xxii; new visibility in, 22; ontology of, 10, 183; post-Fordist, x–xi, xv–xvi; postmodern, 16–18, 167, 170–71, 197; power-time in, 222; production of subjectivity in, 2, 195, 265n29; semiotics in, 174; simulation in, 141; social machine of, 68; technologies of time in, 10; technology and, 7–8, 38; temporality of, 102, 200; time and, 8–9, 38; value-time of, 200–1; against visible world, 20; work-model in, 196
Capitalist machine, 167
Capitalization, 177
Cinelanguage, 29–30
Cinelinkage, 27
Cinema: as adventure, of perception, 219–20; Agamben on, 208–9; against apparatus, 268n28; Benjamin on, xv–xvi, 210–14; Bergson on, 82–83; camera, 83; as collective perception, 205, 217; consciousness and, 29; crystallization of time by, 203, 253n2; Deleuze on, 84; division of labor in, 24–25; as flow of images, xx; illusion of movement in, 82; machines of, xvii, xx–xxi; mass character of, 31; micropolitics of, 31; movement, time in, 217; movement in, 98; organizing, 31–32; postwar, 106, 219; production and, 213; production of subjectivities in, 26; spectacle of, 106; subjectivation and, 21; technique, 213; technology, xviii–xix, 82, 94, 217; television versus, 98–99; temporality in, 209; the unconscious and, 211; Vertov on, xvii, 23–24, 31; video and, 212; as war machine, xvii–xviii
Cinema (Deleuze), xiii, xvii
Cinematic communication, 22
Cinematic image, 211–12

Cinematic signs, xx
Cinematograph, 253n4
Cinesensation, 27
Class vision, 23
Coexistence, of past and present, 57–58
Cognition, 6
Cognitive machines, 163
Cognitive psychology, 157–58
Cognitive science, 160
Collective body, 35
Collective perception, xv, 16–17, 199–200, 204–5, 215–17
Color, 88–89, 142
Comic genre, of literature, 207–8
Commercialization, 209–10
Commodity, 28, 190
Communication: cinematic, 22; of computer, human, 191–92; industry, 212; language and, 171–72; with machine, 192; in production of subjectivity, 171; travel and, 25, 97
Complexity, 77
Computer, 130–32, 136–38, 140; client-server architecture of, 189; human and, 191–92; simulation, 152–53; as smart device, 185; software, 187
Computerization, of society, 186
Concept, 2–3, 5–6, 8–10, 155–56
Consciousness, 5, 247n3, 249nn34–35, 266n2; abolition of, 42–43; attention of, 70–71; automation of, 153; brain and, 153–54; cinema and, 29; computer simulation of, 152–53; duration of, 50; of ego, 164–65; intellect versus, 71; memory and, 40–41; multiplicity of, 154; Nietzsche on, 153; perception and, 86–87; planes of, in memory, 120–21; present for, 249n23; real time of, 153–54
Conservation, of time, 57

Consumer, xv–xvi
Contemporary capitalism. *See* Capitalism
Contraction, 62–63, 104
Contraction-memory, 55–56
Contraction-relaxation: of memory, 113; of time, 55–56, 104
Contraction-syntheses, of time, 39–40
Corcorde, the, 267n25
Couchot, Edmond, 111, 114–15, 118–19, 134–35, 258n8
Creation: Bergson on, 75–77; in capitalism, 76; fabrication versus, 75–76; of the new, as truth, 150; power of, 3
Creative force, 18
Creative labor, 36
Creative power, of time, 60
Creativity, 5, 235–36
Crystallization of time, xix–xxi, xxiv; by cinema, 203, 253n2; by electronic, digital technologies, 1; Marx on, 28; by video, digital technologies, 65–67. *See also* Machines that crystallize time
Cultural production, labor market of, 194
Customer, 184–85, 188–93
Cybernetics, 152

Daney, Serge, 99, 102
Death, 14–15
Deleuze, Gilles, x, xii–xiii, xvii–xviii, xx, 15, 230–31; on capitalism, 170; on cinema, 84; on habit, 63–64; on memory, 42; on passive syntheses, 147; on postwar cinema, 106; on pure perception, 85; on subjectivity, 64; *A Thousand Plateaus* by, 172, 176–78, 195–97; on time, 61–62
Desire, 172–74

Destruction of Reason, The (Lukács), xii
Deterritorialization, 8–9, 33; capitalist, 102, 174, 177; by capitalist machine, 167; by digital technologies, 161–62; restoring mobility, 131
Dialectic of Duration, The (Bachelard), 240n11
Digital apparatuses, 222–23
Digital recording, 136–37
Digital technology, 13, 39, 73–74, 97, 112–16, 133, 136–37; in America, 187; crystallization of time by, 1, 65–67; deterritorialization by, 161–62; interactivity of, 213–14; production and, 184–85; in reality, 7; in technomics, 186–87
Distraction, 70–71, 219
Divination, 6, 248n19
Division of labor, 24–25
Duchamp, Marcel, xvii, 236
Duration, 14–15, 40; body as, 50; of consciousness, 50; as force, 41; memory and, 50–51; as real time, 84; spatialization of, 137; technology and, 73; of things, 86–87
Duration time, 37–39, 41, 87, 132–33
Dynamic schema, 117–18, 122–23, 146
Dziga Vertov Group, 20

Economy, xiv–xv; American, 186–88; high-tech, 189; in production, of subjectivity, 169–83; services in, 186. *See also* Attention economy; Information economy; Political economy
Editing, 25, 91
Editing tables, 91
Ego, 164–65
Eisenstein, Sergei, 24, 35, 95
Electronic music, 101
Electronic technology. *See* Digital technology

Energy, 15–16, 70, 73–74
Enslavement. *See* Machinic enslavement
Ethics, of vision, 7, 144–45
Event, the, xvii–xviii, 100, 160–62
Eye, 6, 25, 55, 145–46, 149, 156, 256n3. *See also* Vision

Fabrication, 75–77
Factory, 178
Falsehood, 148
Fashion, 182
Feelings, 5
Film drama, 23, 31
Flow: Beuys on, 100; of electricity, 252n1; of images, xx, 81; of knowledge, signs, 174; of light, 88–89; in machines that crystallize time, 153; modulation of, 83; signifying, asignifying, 37, 180; of television, 218; in video, 83, 88–89
Fogel, Robert, 184
Force, 13–15, 21; of acting, 42; active, 145–46; of art, 147; creative, 18; duration as, 41; dynamic schema as, 118, 122; image and, 128; of invention, 77; resultant, 26; sign and, 29, 141–42; time and, 40–41, 61–62, 77; willing-artist as, 150. *See also* Affective force
Foucault, Michel, xxiii
Frame, video, 110
Free action, 196–97
Free image, 146
Future, the, 64, 98, 208

Genealogy, of thought, 5–6
Godard, Jean-Luc, 20, 99
Guattari, Félix, x–xi, xix–xxiii, 16, 132–33, 230–31, 262n7; aesthetic paradigm of, 235–36; on animism, 234; on assemblage, 192; on

Guattari, Félix (*cont.*)
capitalism, xxii, 170, 173–76; on Corcorde, 267n25; on modelization, 183, 195; on production of subjectivity, 265n29; against scientific paradigm, 235; *A Thousand Plateaus* by, 172, 176–78, 195–97

Habermas, Jürgen, 192
Habit, 39–56, 63–64, 254n15
Hallward, Peter, xvii
Heidegger, Martin, xviii, xxi–xxii
High-tech economy, 189
Hollywood, 24, 216
Honneth, Axel, xv
Horkheimer, Max, xiv
Human: computer and, 191–92; machine and, 30, 180, 189; nature and, 86; simulation and, 140, 143. *See also* Body
Human-television machine, 181

Idealism, 251n46
Ideological reproduction of, 175–76
Image: actual, 59–60, 67–68; artistic faculty selecting, 147; attention and, 71; body and, 45–46, 48, 111; brain and, 50–51; camera and, 82; cinema as flow of, xx; cinematic, 211–12; construction of, 117; as fabric of being, 5; flow of, xx, 81; force and, 128; free, 146; infinity of, 116; interval and, 27–28; in itself, 138; machine and, 8; matter and, 43–44; in memory, 122; movement and, 125; perception and, 47; as place, 144; processing, 92; producing by synthesis, 132; production of, xiv, xx–xxi, 19–20, 53–55, 73–74, 113–29, 211; in pure perception, 43–44, 85; raw perception and, 124; reality and, 11; regime of, 101; as reification of the visible, 27–28; relationships of, 117–18, 125; representation and, 117–19; schema and, 123–24, 126–29, 134; society of, 45; sound and, 33, 114, 136; space of, 133–34; synthetic, 109–12, 133–34; tactile, 111, 220; technology and, xxi; television, 217–19; as thing, 105; of thought, 29; time and, 3–4, 70; video, 67, 83, 85, 89–90, 99, 133, 220. *See also* Matter-image; Memory-image; Perception-image; Virtual image; World image
Image-matter, 3, 45, 47, 82, 113
Imagination, 132, 135–36, 147, 149
Imitation, 65–66
Immaterial labor, ix–x, xix, xxiv
Industrial labor, 185
Industrial Revolution, 4
Information, 104
Information economy, 17, 172–73, 175, 182; as apparatus of capture, 185; customer in, 184–85, 188–93; exploitation of free action in, 196–97; on industrial labor, 185; innovation in, 190; machines that crystallize time in, 176; organization, dynamics of, 184; production of services in, 193; production of subjectivity in, 183, 192; service economy versus, 264n21; socioeconomic description of, 183; software of, 193
Innovation, 190
Intellect, 2–3, 71–75, 134, 250n42
Intellectual labor, 17, 32, 36, 56, 114, 116–17, 121–24, 126–29, 131–34, 222, 240n17
Interactive television, 100
Interactivity, 111, 213–14
Interface, brain as, 46–47

Intervals, 27–28
Intuition, 3, 74–75
Invention, 77, 123
Investment, 186–87
"I see," 22

James, William, xxiii
Jetztzeit, xv–xvi, 201, 203
Judgment, 143, 163–64

Kafka, Franz, 25
Kasparov, Garry, 261*n*22
Kino-eye, xvii–xviii, 19; connecting proletariat through, 27; "I see" of, 22; as machine, xviii, 21, 24, 28; perception of, 24, 30, 241*n*27; Vertov on, 24–26, 28, 32–36, 90, 96, 105; vision of, 25–26; visual thinking of, 29
Kinoki movement, 19–21, 24–25, 32–35
Knowledge, 157, 159–62, 163, 174

Labor: becoming-active of, 213; capitalism exploiting, 173; commodity and, 28; creative, 36; division of, 24–25; immaterial, ix, xix, xxiv; industrial, 185; intellectual, 56, 114, 116–17, 121–24, 126–28, 131–34, 240*n*17; living, 9, 17, 171; manual, intellectual, xv–xvi, 17, 32, 36, 222; market, 194; of memory, 91; perception and, 1–2; production of subjectivity, 195; surplus value and, 196; wage, 172
Language, 5, 18, 72–73, 134, 152, 250*n*38; communication and, 171–72; of man, 192; in production of subjectivity, 171
Laughter, 207–8
Lawrence, D. H., 257*n*6
Lazzarato, Maurizio, ix–xx, xxiii–xxiv, 227–28

Leibniz, Gottfried, xii–xiv, xviii, 15–16, 21
Life, 174–75
Life Caught Unawares, 26–27
Light, 47, 54, 85, 88–89, 113–16, 134
Live technology, 100
Living bodies, 79
Living labor, 9, 17, 171
Logic, 158–59
Lukács, György, xii

Machine: capitalist, 167; of cinema, xvii, xx–xxi; cognitive, 163; communication with, 192; human and, 30, 180, 189; human-television, 181; image and, 8; kino-eye as, xviii, 21, 24, 28; man and, 176–77; of market, xxv; power of, 131; seeing, thinking, 21; social, 34, 68, 192; soul and, 4; technological, xxii, 68; triumphing over mechanism, 70–79, 130, 133; worker and, 177. *See also* Automation; Video; War machine
Machines that crystallize time, xxv, 9, 11, 15–16, 21, 52, 68–69, 132–33; in bifurcation of time, 180–81; in capitalism, 175, 203; flow in, 153; in information economy, 176; in production of subjectivity, 169–70; semiotization by, 174; subjectivation by, 191
Machinic assemblages, xxii, 100, 160, 181
Machinic enslavement, x–xi, xix, xxii–xxiii, 177–81, 191, 197–98
Machinic subjectivity, 191
Malevich, Kazimir, 117, 246*n*1
Man, 140, 176–77, 192. *See also* Human
Manual labor, xv–xvi, 17, 32, 36, 222
Marx, Karl: Bergson and, 229; on crystallization of time, 28; on

Marx, Karl (*cont.*)
living labor, 9, 17, 171; methodology of, 229; on subjectivity, 170; on subject-object relations, 230; on work, 195–96
Marxism, 231, 233, 235; methodology of, 171; political ontology of, ix–x; on production of value, 8–9; on subjectivity, of workers, 170; theory of value, 1
Mass character, of cinema, 31
Mass customization, 187–88, 191
Masses, the, 204–7, 216, 225
Mass reproduction, 203
Materialism, 112, 251*n*46
Materiality, subjectivity and, 223
Material synthesis, 39–40, 42, 44, 65–69
Matrix-image, 109–11
Matter, 9, 248*n*10; consciousness and, 86–87; image and, 43–44; processed by video machine, 45; undulation of, 83; in video, 131. *See also* Image-matter; Perception-matter; Time-matter
Matter and Memory (Bergson), xv, 38, 120–21, 228
Matter-energy, 88
Matter-image, 64–65
McCraken, Edward, 189–90
McLuhan, Marshall, 11–12, 83
Mechanical mysticism, 4, 77
Mechanism, 70–79, 130, 133
Melitopoulos, Angela, 228, 255*nn*16–17
Memory, 14–15; automatic versus attentive recognition, 50–52; Bergson on, 41–42, 50–53, 176, 201; body and, 50–53; consciousness and, 40–41; contraction by, 104; contraction-relaxation of, 113; duration and, 50–51; dynamic scheme in, 123; as extension of past, into present, 103; gap in, 40; habit and, 39–56, 64, 254*n*15; images in, 122; involuntary, 200; judgment and, 164; labor of, 91; montage and, 90–95; ontological, 11, 56–65; perception and, 50, 94, 133; planes of consciousness in, 120–21; psychological, 56; social, 102; as spiritual synthesis, 44; television and, 103–4; time and, 59–60; virtual, 68, 202; as virtuality, 53. *See also* Contraction-memory; Pure memory; Recollection-memory
Memory-image, 51
Metacinema, xii–xiii
Micropolitics, xi–xii, 31, 35
Modelization, 16, 183, 195, 263*n*12
Money, xi. *See also* Capital
Montage, 25, 90–95
Motors, of affective energy, 73–74
Movement: body and, 125; brain and, 46–47, 50–51; camera and, 28; in cinema, 98; illusion of, 82; image and, 125; perception of, 63; of representation, 118; of time, 94–95, 98, 217; in video, 93. *See also* Kinoki movement
Multiplicity, 154, 156–57, 164–67, 257*n*7
Mysticism, mechanical, 4

Natural perception, 42
Nature, human and, 86
Negative dialectics, xv
Negative ontology, 4–5
Negri, Antonio, xxiv–xxv, 102
Neoliberalism, xi, xiii–xiv
New barbarism, 202–3, 226, 232–33
Newcomen, Thomas, 76
New economy. *See* Information economy; Technomics
Nietzsche, Friedrich, xii–xiii, 1, 25, 226; on action, 162–63; on

appearance, 144; on artists, 235; Bergson and, x, 13–14, 16, 229–30; on the body, 5–7, 164–67; on concept, reality, 10; on consciousness, 153; on eye, 145–46, 149, 156; on force of art, 147; on forces, signs and, 141–42; on logic, 158–59; metaphysical thesis of, 243n7; on percept-concept relation, 155–56; on simulation, 142–43; on thinking, 155–56, 162–63; on third eye, 151–52; on thought, 155–58, 163; on thought, action, 152, 154–55; on truth, 159; on truth, the real, 150; virtual technology and, 139–40; on vision, 145–46, 259n8; on will to power, 60, 140, 145, 148, 150, 244n13

Notes of a Film Director (Eisenstein), 95

Nothingness, being and, 4–5

Novel, the, 207–9

Object, subject and, 44–45, 148, 163, 228–30

Onto-aesthetics, x, xvii–xviii, xxiv

Ontological consistency, of video, 98–99

Ontological memory, 11, 56–65

Ontology: of Bergson, 10, 182; of capital, 10; of contemporary capitalism, 10, 183; negative, 4–5; perception in, 45; political, ix–x, xiv, xvii, 231–32; positive, 4–5; reality, concept in, 8; of television, 103–4; temporal, 1, 197; of time, 62, 202

Optical model, of production of images, 53–55, 211

Organism, 48, 78–79, 164–65

Original time, 8

Paik, Nam June, 12–13, 39, 84, 86–89, 92, 94, 99–100, 110

Pasolini, Pier Paolo, 230

Passive habits, 63–64
Passive syntheses, 147
Passivity, 41–42, 220–21
Past, absolute, 206
Past, the, 56–58, 98, 104, 201–2. *See also* Memory
Percept, 2–3, 155–56
Perception: action and, 69; adventure of, 219–20; ancient Greeks on, 2; apprehension of, 52–53; attentive, 52–53; Bergson on, 2–3, 63, 87–88, 146, 148; cognition and, 6; collective, xv, 16–17, 199–200, 204–5, 215–17; collectivization, mechanization of, xv; consciousness and, 86–87; contraction in, 62–63; image and, 47; as instrument, of acceleration, 96; of kino-eye, 24, 30, 241n27; knowledge and, 160–61; labor and, 1–2; limits of, 89–90; memory and, 50, 94, 133; of movement, 63; natural, 42; in ontology, 45; pure, xii–xiii, 11; raw, 117, 124; recollection and, 60; seeing eye, 6; sensation over, 48, 52–53; sensory, 147–48; subject, object and, 44–45; subjectivation and, 145; by technology, 66; in things, 105; time, affect and, 65; understanding and, 3, 155; video recording and, 91; of virtual technology, 144. *See also* Pure perception
Perception-image, xvii, 40, 105
Perception-matter, 85–86, 92, 109
Perceptive synthesis, 62
Photogram, 82–83
Photography, xx, 82
Pixel, 109–10
Political action, xxiv
Political economy, 9–10, 18, 183, 190–91
Political onto-aesthetics, x

Political ontology, ix–x, xiv, xvii, 231–32
Politics, of eye, 146
Positive ontology, 4–5
Post-Fordism, x–xii, xv–xvi, xxiv
Post-media era, 183
Postmodernism, xxiv–xxv, 10, 16–18, 167, 170–71, 197, 223
Postwar cinema, 106, 219
Power: of acting, 41; appearance and, 144; of digital technologies, 7; formation, 176; of machine, 131; production and, xxiii; television, as apparatus of, 94, 101; time and, 102, 207. *See also* Will to power
Power signs, xxiii
Power-time, 203, 221–22, 226
Present, the, 56–58, 103, 201–2, 249*n*23
Processing, 90–95
Producer, xv–xvi, 190
Production: activity of, 121; artistic, 193–94; cinema and, 213; continuity, of processes of, 182–83; cultural, 194; customer and, 184; digital technologies and, 184–85; of factory, 178; by film industry, 31; of images, xiv, xx–xxi, 19–20, 53–55, 73–74, 113–29, 211; in machinic enslavement, 179; mass customization, 187–88; means of, 175; nonstandardized mode of, 182; in postmodern capitalism, 170–71; power and, xxiii; of services, 193; of value, 8–9, 185
Production, of subjectivity, xi, 16–17, 68–69; in capitalism, 2, 195, 265*n*29; in cinema, 26; communication in, 171; in economy, 169–83; in information economy, 183, 192; labor and, 195; by machines that crystallize time, 169–70; television in, 103
Profit, 186–89

Projection room, 25
Proletariat, 23, 27
Psychological memory, 56
Psychology, 48, 157–58
Public sphere, 23
Pure memory, 56–65
Pure perception, xii–xiii, 11; bodies and, 46; Deleuze on, 85; as image, 85; image-matter in, 45; through material, spiritual syntheses, 44; natural perception versus, 42; through technology, 86; of video, 45, 66, 211

Qualitative time, 10–11

Radio-ear, 34
Raw perception, 117, 124
Real, the, 140–42, 145, 150, 174, 224–25
Reality, 6–11
Real time, 66–67, 260*n*17; of consciousness, 153–54; duration as, 84; splitting of time, 96–105; in video technology, 151
Reception, 219
Recognition, 40, 50–52, 55
Recollection, 60, 121–22
Recollection-memory, 56
Recording, digital, 136–37
Redstone, Sumner, 193
Reification, of the visible, 27–28
Representation, 45, 95, 98–99, 106–7, 117–19
Reproduction: ideological, 175–76; mass, 203; technological, xv, 200
Reversibility, of producer and consumer, xv–xvi
Revolution, 216
Riefenstahl, Leni, 35, 216
Rossellini, Roberto, 101
Russia. *See* Kinoki movement; Soviet Union

Sauvagnargues, Anne, xiii
Schizoanalytic modelization, 183
Science, 74–75, 235, 251*n*43
Screenplay, 24–25
Seeing eye, 6
Seeing machines, 21
Self-awareness, 143
Semantics, xxiii
Semiotics, 141–42, 230–31; asignifying, xxii–xxiv, 176; in contemporary capitalism, 174; signifying, xxii–xxiii; time of life and, 2
Semiotization: by machines that crystallize time, 174; the real and, 141–42; simulation technologies in, 148–49
Sensation, 14, 48–49, 52–53
Sensible, 5
Sensory-motor activity, of body, 51–52, 91
Sensory perception, 147–48
Service economy, 264*n*21
Services, 186, 193
Shooting, 94–95
Signifier, 192
Signifying flows, 37, 180
Signifying semiotics, xxii–xxiii
Signifying signs, xix
Signs, xix–xx; flow of, 174; forces and, 29, 141–42; in post-Fordism, xxiv; power, xxiii; time, 256*n*30
Simondon, Gilbert, 73
Simulation, 93; in capitalism, 141; computer, of consciousness, 152–53; human and, 140, 143; Nietzsche on, 142–43; technology of, 110, 112–13, 130, 132, 135, 148–49, 161
Situation, 100
Situationists, 193
Social, the, 224–25
Social machines, 34, 68, 192
Social memory, 102

Social practices, 166
Social subjection, xxiii
Society: body in, 165–66; computerization of, 186; contemporary capitalism exploiting, 173; of the image, 45; subjectivity and, 178
Software, 187, 193
Soul, 4
Sound, image and, 33, 114, 136
Soviet revolution, 20–21
Soviet Union, 32, 34
Space, 12, 133–34
Space-time, 101
Spectacle, xvii–xviii, 19–20, 105–7, 218
Spectator, 220–21
Speed, 47, 161
Spinoza, Baruch, 21, 231–32
Spirit, 6–7
Spirituality, 4
Spiritual synthesis, 40, 42, 44, 65–69
State, the, 98–99
Striated capital, 197
Subject: of "I see," 22; object and, 44–45, 148, 162, 228–30
Subjection, 177–80, 197–98
Subjectivation, 5, 16, 21, 145, 150, 180–82, 191
Subjectivity, x–xi, xviii–xix, 244*n*18; bifurcations, experimentation of, xxv; Deleuze on, 64; as internal to perception, time and affect, 65; machinic, 191; manipulation of, xxiv; materiality and, 223; society and, 178; time and, 11, 35; work and, 170–71; of workers, 170. *See also* Production, of subjectivity
Subject-objects, 112
Supersensible, 5
Surplus value, 196
Symbols, 160
Synapses, 179–80

Syntheses: of body, 164–66; of body and memory, 53; images produced by, 132; material, 39–40, 42, 44, 65–69; passive, 147; perceptive, 62; spiritual, 40, 42, 44, 65–69; temporal, 55; of time, 4, 12–13, 62, 64
Synthetic image, 109–12, 133–34

Tactile image, 111, 220
Tactile optics, 212
Tarde, Gabriel, xii–xiv, 119–20
Techne, xxi–xxii
Technological machine, xxii, 68
Technological reproduction, xv, 200
Technology: absorbing, distracting attention of consciousness, 70–71; automism, 158; capitalism and, 7–8, 10, 38; cinema, xviii–xix, 82, 94, 217; duration and, 73; image and, xxi; live, 100; miniaturization of, 30–31; perception by, 66; post-Fordist, xi; in pure memory, 57; pure perception through, 86; of self, artistic practices as, xxiv; of simulation, 110, 112–13, 130, 132, 135, 148–49, 161; simulation by, 93; theories of, 54; of time, 10, 61, 65, 161, 203; virtual, 111, 139–40; of vision, 144–45. *See also* Digital technology; Machine; Video
Technology paradox, the, 187
Technomics, 184, 186–87, 189–90, 192–93, 263n20. *See also* Information economy
Teleological action, 71–72, 75
Television, xxi, 33–34, 68, 84; as apparatus of power, 94, 101; cinema versus, 98–99; control by, 102–3; destroying the public, 219; Eisenstein on, 95; existence of, 102; flow of, 218; images, 217–19; interactive, 100; memory and, 103–4; ontological foundation of, 103–4; as passive reception device, 185; the past in, 104; in production of subjectivity, 103; spectacle, 218; spectator, 220–21; as time, 104–5; video versus, 220; viewers, 179–81
Temporality: Bergson on, 87–88; of body, 49; capitalist, 102, 200; in cinema, 209; pure present, pure past in, 57; of video, 95; virtual, 111
Temporal model, of production of images, 53–54
Temporal ontology, 1, 197
Temporal synthesis, 55
Thinking, 155–56, 162–63. *See also* Thought
Thinking machines, 21
Thought, 5–6; action and, 152, 154–55; body and, 164; forms of, 156; image of, 29; knowledge and, 163; multiplicity in, 156–57; Nietzsche on, 155–58, 163
Thousand Plateaus, A (Deleuze and Guattari), 172, 176–78, 195–97
Time: abstract, 60–61, 260n17; acceleration of, 96–97; accumulation-conservation of, 50; Bakhtin on, 206; conservation of, 57; contraction-relaxation of, 55–56, 104; contraction-syntheses of, 39–40; creative power of, 60; deterritorialization of, 8–9; double foundation of, 201–2; double grounded, in ontological memory, 57; duration, 37–39, 41, 87, 132–33; of the event, 161–62; force and, 40–41, 61–62, 77; hierarchy of, 206–7; image and, 3–4, 70; measurement and, 8, 184; memory and, 59–60; movement of, 94–95, 98, 217; ontology of, 62, 202; passing, in pure memory, 58–59;

perception, affect and, 65; power and, 102, 207; qualitative, 10–11; real, 66–67; real versus abstract, 60–61; representation of, 99; sensation in, 14; signs, 256n30; space and, 12, 101; splitting, 58–60, 96–105, 210; in subjectivity, 65; subjectivity and, 11, 35; syntheses of, 4, 12–13, 62, 64; technologies, in capitalism, 10, 38; technology of, 61, 65, 161, 203; television as, 104–5; video and, 12–13, 39, 66–67, 88, 93–94, 97–98, 219. *See also* Crystallization of time; Machines that crystallize time; Power-time; Value-time; Whatever-time

Time-image, 49, 93

Time-matter, xix, 3, 45, 51, 92, 113, 136, 218

Time of life, 1–2

Toscano, Alberto, xv

Travel, 25, 97

Truth, 150, 159

Two Sources of Morality and Religion (Bergson), xiv, 242n2

Unconscious, the, 157–60, 166, 210–11
Understanding, perception and, 3, 155
Universe, xii–xiii
Unthought, the, 6

Valorization, 170, 174–75
Value, 1, 8–9, 185, 196
Value-time, 200–2, 221
Varela, Francisco, 159–60
Vertov, Dziga, 19–23; on camera, 81; on cinema, xvii, 23–24, 31; on image, as reification of the visible, 27–28; on kino-eye, 24–26, 28, 32–36, 90, 96, 105; on social machine, 34
Vibration, 86–87

Video, xvii–xviii; brain and, 66, 88; cinema and, 212; crystallizing time, 65–67; event in, 100, 162; flow in, 83, 88–89; frame, 110; images, 67, 83, 85, 89–90, 99, 133, 220; imitating time, 39; imitation by, 65–66; input, output, 99; as machinic assemblage, of situations, 100; as material, spiritual synthesis, 65–69; matter in, 131; as modulation of flows, 83; movement in, 93; ontological consistency of, 98–99; perception and, 91; pure perception of, 45, 66, 211; real time in, 151; sensory-motor function and, 91; space-time blocks of, 101; state and, 98–99; subject, object and, 228–29; technology, xviii–xxi, 12, 37–38, 65–66, 81–83, 94, 110; television versus, 220; temporality of, 95; time and, 12–13, 66–67, 88, 93–94, 97–98, 219; virtuality of, actualized in synthetic images, 109–12

Video art, xxiv, 12, 101
Video camera, xxi, 12, 67, 92–93, 211
Video machine, xxi, 45
Viewers, 179–81, 286n12
Viola, Bill, 89, 93, 133–34
Virilio, Paul, xix, 96–97, 161
Virtual, the: the actual and, 56–57, 60, 71–72, 105–6, 126, 129, 244n17; concept of, 129–38; vision and, 146. *See also* Actual-virtual circuit
Virtual image, 59–60, 67–68, 101, 106, 129–30, 144
Virtuality, 21, 23, 35, 40, 52–53, 109–12
Virtual memory, 68, 202. *See also* Pure memory
Virtual technology, 111, 139–40; as false world, 150; perception of, 144; truth and, 150

Virtual temporality, 111
Visibility, 20, 22, 220
Visible, the, 27–28
Vision: Bergson on, 259n9; class, 23; ethics of, 7, 144–45; of kino-eye, 25–26; Nietzsche on, 145–46, 259n8; in optical model, 54–55; technology of, 144–45; in temporal synthesis, 55; the virtual and, 146
Visual impression, 134
Visual thinking, 29
Vitalism, xi–xiii

Wage labor, 172
War machines, xvii–xviii, 33
Wave, 109
Wealth production, 174

Weissberg, Jean-Louis, 97
Whatever-subjectivity, 195–96. *See also* Subjectivity
Whatever-time, 8–9, 222
Will, 2, 78, 141–42
Willing-artist, 150
Will to power, 6, 13–14, 60, 140, 148, 150, 244n13
Will to truth, 145
Words, 29–30, 72–73, 160
Work: Marx on, 195–96; as motor cause, 195; refusal of, 222; subjectivity and, 170–71
Worker, 170, 177, 221. *See also* Labor
Work-model, 195–96
World, as flow, of images, 81
World image, 20, 43, 84

COLUMBIA THEMES IN PHILOSOPHY,
SOCIAL CRITICISM, AND THE ARTS

LYDIA GOEHR AND GREGG M. HOROWITZ, EDITORS

John T. Hamilton, *Music, Madness, and the Unworking of Language*

Stefan Jonsson, A Brief History of the Masses: Three Revolutions

Richard Eldridge, *Life, Literature, and Modernity*

Janet Wolff, The Aesthetics of Uncertainty

Lydia Goehr, *Elective Affinities: Musical Essays on the History of Aesthetic Theory*

Christoph Menke, *Tragic Play: Irony and Theater from Sophocles to Beckett*, translated by James Phillips

György Lukács, *Soul and Form*, translated by Anna Bostock and edited by John T. Sanders and Katie Terezakis with an introduction by Judith Butler

Joseph Margolis, *The Cultural Space of the Arts and the Infelicities of Reductionism*

Herbert Molderings, *Art as Experiment: Duchamp and the Aesthetics of Chance, Creativity, and Convention*

Whitney Davis, *Queer Beauty: Sexuality and Aesthetics from Winckelmann to Freud and Beyond*

Gail Day, *Dialectical Passions: Negation in Postwar Art Theory*

Ewa Płonowska Ziarek, *Feminist Aesthetics and the Politics of Modernism*

Gerhard Richter, *Afterness: Figures of Following in Modern Thought and Aesthetics*

Boris Groys, *Under Suspicion: A Phenomenology of the Media*, translated by Carsten Strathausen

Michael Kelly, *A Hunger for Aesthetics: Enacting the Demands of Art*

Stefan Jonsson, *Crowds and Democracy: The Idea and Image of the Masses from Revolution to Fascism*

Elaine P. Miller, *Head Cases: Julia Kristeva on Philosophy and Art in Depressed Times*

Lutz Koepnick, *On Slowness: Toward an Aesthetic of Radical Contemporaneity*

John Roberts, *Photography and Its Violations*

Hermann Kappelhoff, *The Politics and Poetics of Cinematic Realism*

Cecilia Sjöholm, *Doing Aesthetics with Arendt: How to See Things*

Owen Hulatt, *Adorno's Theory of Philosophical and Aesthetic Truth: Texture and Performance*

James A. Steintrager, *The Autonomy of Pleasure: Libertines, License, and Sexual Revolution*

Paolo D'Angelo, *Sprezzatura: Concealing the Effort of Art from Aristotle to Duchamp*

Fred Evans, *Public Art and the Fragility of Democracy: An Essay in Political Aesthetics*

GPSR Authorized Representative: Easy Access System Europe, Mustamäe tee
50, 10621 Tallinn, Estonia, gpsr.requests@easproject.com

www.ingramcontent.com/pod-product-compliance
Lightning Source LLC
Chambersburg PA
CBHW021937290426
44108CB00012B/868